continued on back

Spatial Statistics

Spatial Statistics

BRIAN D. RIPLEY

Reader in Statistics
Imperial College
University of London

JOHN WILEY & SONS

New York • Chichester • Brisbane • Toronto • Singapore

Library of Congress Cataloging in Publication Data:

Ripley, Brian D   1952-
    Spatial statistics.

    (Wiley series in probability and mathematical
statistics: applied probability & math section)
    Bibliography: p.
    Includes index.
    1. Spatial analysis (Statistics) I. Title.

QA278.2.R56   519.5'37   80-26104
ISBN 0-471-08367-4

Printed in the United States of America

10 9 8 7 6 5 4 3 2

Preface

This is a guide to the analysis of spatial data. Spatially arranged measurements and spatial patterns occur in a surprisingly wide variety of scientific disciplines. The origins of human life link studies of the evolution of galaxies, the structure of biological cells, and settlement patterns in archeology. Ecologists study the interactions among plants and animals. Foresters and agriculturalists need to investigate plant competition and account for soil variations in their experiments. The estimation of rainfall and of ore and petroleum reserves is of prime economic importance. Rocks, metals, and tissue and blood cells are all studied at a microscopic level. The aim of this book is to bring together the abundance of recent research in many fields into the analysis of spatial data and to make practically available the methods made possible by the computer revolution.

The emphasis throughout is on looking at data. Each chapter is devoted to a particular class of problems and a data format. The two longest and most important are on *smoothing and interpolation* (producing contour maps, estimating rainfall or petroleum reserves) and on *mapped point patterns* (trees, towns, galaxies, birds' nests). Shorter chapters cover:

The regional variables of economic and human geography.

Spatially arranged experiments.

Quadrat counts.

Sampling a spatially correlated variable.

Sampling plants and animals and testing their patterns.

The final chapter looks briefly at the use of image analyzers to investigate complex spatial patterns, and stereology: how to gain information on three-dimensional structures from linear or planar sections. Some emphasis is placed on going beyond simple tests to detect "nonrandom" patterns as well as on fitting explanatory models to data. Some general families of models are discussed, but the reader is urged to find or invent models that reflect the theories of his or her own discipline, such as central place theory for town locations. The techniques presented are designed for both of John Tukey's divisions of exploratory and confirmatory data analysis.

The level of mathematical difficulty varies considerably. The formal prerequisites are few: matrix algebra, some probability and statistics, and basic topology in parts of Chapter 9. An acquaintance with time series analysis would be helpful, especially for Chapter 5. I have tried to confine the formal mathematics to the essential minimum. Mathematically able readers will be able to find their fill in the references. It is perhaps inevitable that some of the mathematical justifications are far deeper than is the practical import of the results. But beware. There is much appealing but incorrect mathematics in the spatial literature, and some of the subtlest arguments are used to discover undesirable properties of simple procedures. I recommend readers who find the going tough to skip ahead to the examples before seriously tackling the theory.

Computers, especially computer graphics, are an essential tool in spatial statistics. Useful data sets are too large and most of the methods too tedious for hand calculation to be contemplated. Even data collection is being increasingly automated. The worked examples were analyzed at an interactive graphics terminal by FORTRAN programs running on Imperial College's CDC 6500/Cyber 174 system. Unfortunately, the reader cannot follow my decisions as I rotated plots, investigated contour levels, and altered smoothing parameters. There is no substitute for experience at a computer terminal using one's own data. Therefore, it was a difficult decision not to include programs. There was at the time of writing no agreed-upon standard for computer graphics, and the availability of plotting and other utility operations varied widely. The choice of language was also debatable. I could only use interactive graphics from FORTRAN, whereas microcomputers were becoming available with BASIC or PASCAL. Hints on algorithms and computation are included.

The bibliography is the only example I know of that attempts a comprehensive coverage of the spatial literature. It contains not only references to the theory and methods, but a large number of accounts of applications in many disciplines as well. Guides to the literature are given at the end of several chapters and sections.

B. D. RIPLEY

London
March 1981

Acknowledgments

This book was written during two periods of leave visiting the Department of Statistics, Princeton University and the Afdeling for Teoretisk Statistik, Aarhus University, Denmark. My stay at Princeton was supported by contract EI-78-5-01-6540 with the U.S. Energy Information Administration. I am grateful to Geof Watson and Ole Barndorff-Nielsen for their interest and encouragement.

Most of the figures were computer-drawn on 35-mm microfilm at the University of London Computer Centre, using procedures set up by Imperial College Computer Centre. The perspective plotting routines were developed jointly with Dan Moore. The Dirichlet tessellations and Delaunay triangulation were drawn by the program TILE of Peter Green and Robin Sibson.

Karen Byth read through the manuscript and removed many errors. I would appreciate being informed of any remaining errors and of work I have missed.

B. D. R.

Contents

Spatial Statistics

CHAPTER 1

Introduction

1.1 WHY SPATIAL STATISTICS?

Men have been drawing maps and so studying spatial patterns for millenia, yet the need to reduce such information to numbers is rather recent. The human eye and brain form a marvelous mechanism with which to analyze and recognize patterns, yet they are subjective, likely to tire, and so to err. The explosion in computing power available to the average researcher now makes it possible to do routinely the intricate computations needed to explore complex spatial patterns.

One sense of the word "statistics" is a collection of numbers, and spatial statistics includes "spatial data analysis," the reduction of spatial patterns to a few clear and useful summaries. But statistics goes beyond this into what John Tukey has called "confirmatory data analysis," in which these summaries are compared with what might be expected from theories of how the pattern might have originated and developed. Consider, for example, Figure 1.1*a*, which is a map of trees in a rectangular plot. Figure 1.1*b* shows a summary of these data as a graph, together with confidence limits for the sort of graph we would get if each tree had been placed at random in the plot, without any reference to the positions of the rest of the trees. This example shows one of the characteristic features of the subject. There are so many different types of spatial patterns that we need to summarize the data in one or more graphs rather than by single numbers, such as the mean and standard deviation of classical statistics.

Almost invariably we will have only a single example of a particular pattern rather than the many replications of measurements found in the experimental sciences. To get some idea of the variability of such data, we are forced to make some assumption of stationarity of the underlying mechanism that generated the pattern. Such an assumption has often been disputed, particularly in the geographic literature. Its validity may depend on the questions being asked. For instance, if we are looking at

1

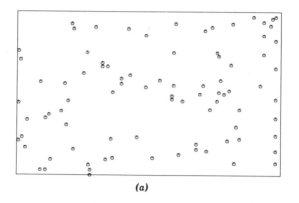

(a)

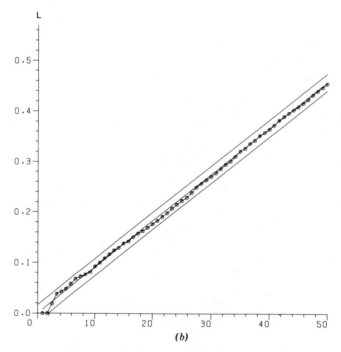

(b)

Fig. 1.1 (a) Point patterns of trees. (b) Summary of data with a 95% confidence band. See Figure 8.6 for further details.

population density we may wish to know whether we need to invoke the topography (which might suggest nonstationarity) to explain the observed variations in density. Patterns that vary in a systematic way from place to place are called *heterogeneous* (opposite *homogeneous*). But we might be studying the grouping of houses that we might expect to *interact*, either *clustering* together because of human gregariousness or *inhibiting* where houses need to be close to sufficient land. Patterns can also exhibit preferred directions, called *anisotropy* (opposite *isotropy*). For example, forests that were originally planted in rows may show directionality in the crowns of the trees (Ford, 1976). We will assume that the data have been subdivided into sufficiently small units or that they have had obvious trends removed to permit us, where necessary, to invoke homogeneity or isotropy.

1.2 TYPES OF DATA

The basic subdivision of this volume is by the type of data to be analyzed. The tree positions given in Figure 1.1*a* are an example of a *point pattern*. Other examples are the locations of birds' nests, of imperfections in metals or rocks, galaxies, towns, and earthquakes. Of course, none of these is actually a point, but in each case the sizes of the objects are so small compared with the distances between them that their size may be ignored. (Sometimes size is an important explanatory variable associated with a point. For example, we might expect the area of the hinterland of a town to depend on its population size.) Maps of point patterns are discussed in Chapter 8.

Sometimes points are so numerous that complete mapping would be an unjustified effort (consider clover plants in a grassland). Two methods of sampling such point patterns are discussed in Chapters 6 and 7. In Chapter 6 we consider methods based on taking sample areas, called *quadrats*, and counting objects within each, whereas in Chapter 7 the methods are based on measuring distances to or between objects. Chapter 7 also deals with two cases in which complete mapping is either uneconomical or impossible; trees in a dense forest and animal populations such as deer and moorland grouse (game birds).

Many variables that were originally point patterns are recorded as regional totals, such as census information. If these regions are genuinely distinct, we may wish to test for correlation between the regional statistics, taking account of the connections between the regions measured by, say, the lengths of the common borders (if any) or freight costs between them.

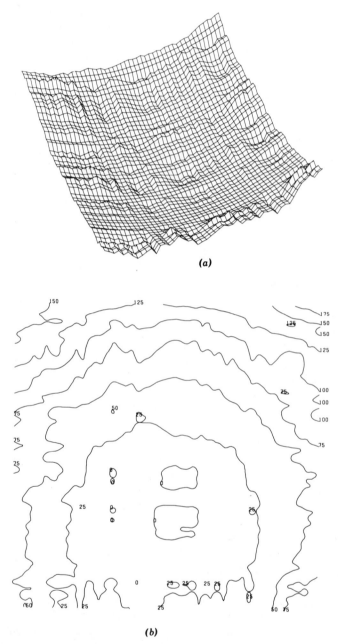

(a)

(b)

Fig. 1.2 (*a*), (*b*) A simulated surface. (*c*), (*d*) Two reconstructions from the sample points indicated as circles.

4

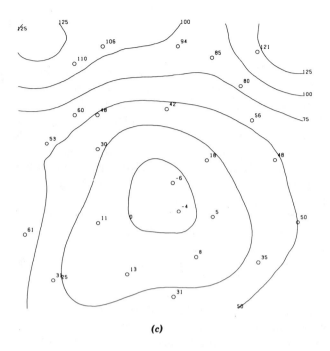

(c)

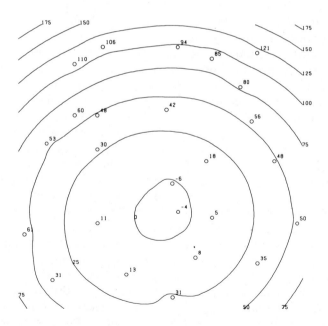

(d)

Fig. 1.2 (*continued*)

5

Summary measures for what is known as "spatial autocorrelation" are discussed in Section 5.4. They are particularly useful when applied to the residuals from the regression of one regional statistic on others.

Where the regions are small administrative units we might wish to smooth the data to produce a map of population density, average income, or similar variable. This problem of reconstructing a surface from irregularly spaced sample points is common; all topographical maps are prepared from such data, as is rainfall information. Geologists, oil prospectors, and mining engineers all have to reconstruct facets of an underground pattern such as the volume and average grade of ore in various parts of a mine, using spatially arranged samples. Such problems are considered in Chapter 4. Figure 1.2 illustrates a surface and two reconstructions.

Usually the locations of the sample points are fixed from other considerations, but in Chapter 3 we consider how sample points should be chosen to give the best estimate of the average level of a surface.

Data arranged on a rectangular grid are not as common as might be expected by analogy with time-series theory. They seem to arise only from man's experiments, either where he has deliberately sampled systematically or from agricultural field trials in which a field has been divided into rectangular parcels. Clearly, we would expect neighboring plots to have similar fertility and hence that the yields would be spatially autocorrelated. We show in Chapter 5 how such data might be analyzed.

The least explored class of patterns are those of two or more phases forming a mosaic. Patterns of vegetation provide two-dimensional examples, but most of the interest is in three dimensions, in bone and tissue and rock grains and pores. Descriptions of patterns such as that shown in Figure 1.3 were facilitated by the invention of image analyzers during the 1960s, these being scanning microscopes connected to computers to analyze the vast amounts of output. Stereology is the theory of reconstructing information on three-dimensional patterns from planar sections (see, for example, Figure 1.3) or linear probes. This area is the subject of Chapter 9.

All the models of the mechanisms that might generate patterns described in the chapters for each type of data are stochastic processes. Chapter 2, on "basic stochastic processes," gives an introduction to what is needed of the mathematical theory, to generic families of models, and to ways in which the computer can be used to experiment with models.

Most of the theory and methods apply equally in two or three dimensions. Where formulas depend on the dimension, only the two-dimensional case is given unless otherwise stated. Planar data are by far the most common; all the examples are planar.

Fig. 1.3   Simplified pore space (black) in a section of smackover carbonate rock.

Spatial Topics Omitted

This volume concentrates on information on location, ignoring the concepts of shape and form reflected in the monographs of Grenander (1976, 1978), Mandelbrot (1977), and Bookstein (1978). Some specialized topics omitted are on the spread of epidemics (Bartholomew, 1973; Mollison, 1977) and percolation theory (Shante and Kilpatrick, 1971; Welsh, 1977; Smythe and Wierman, 1978). Each of these references is more concerned with mathematical modeling than with analyzing data.

Little attention is given here to space–time problems. Many of the same methods can be used, but adequate data seem rare (earthquake occurrences being an exception). Often the best way to deal with space–time data is to compare the maps in successive time periods. Another generalization is to multitype problems in which the objects are of different types or where two or more patterns or surfaces are to be related. Again, the extension of many of the methods is simple. Whenever a pair of points is considered, take one from each of the two patterns or surfaces. If three or more surfaces or patterns are considered, take them in pairs. In general, the theory of multitype procedures is not satisfactory and there are few examples of its use. Pielou (1977) gives examples of some of the methods of "classical" statistics used on these problems.

More information on applications in specific disciplines may be found in:

Animal ecology	Southwood (1978)
Archeology	Hodder and Orton (1976)
Geography	Bartels and Ketellapper (1979)
	Bennett (1979)
	Berry and Marble (1968)
	Cliff and Ord (1973)
	Getis and Boots (1978)
	Haggett et al. (1977)
	Rayner (1971)
	Rogers (1974)
Geology	Davis (1973)
Mining	David (1977)
	Guarascio et al. (1976)
	Journel and Huijbregts (1978)
	Matheron (1965, 1967a)
Plant ecology	Greig-Smith (1964)
	Kershaw (1973)
	Patil et al. (1971)
	Pielou (1977)

CHAPTER 2

Basic Stochastic Processes

This chapter assumes a basic knowledge of probability theory and sets up some of the background of the models and methods used in later chapters. Section 2.4 is more mathematical and is not necessary for an understanding of the rest of the material (although its ideas are used in Sections 5.2 and 8.4).

2.1 DEFINITIONS

A *stochastic process* is a collection of random variables $\{Z(t)|t \in T\}$ indexed by a set T. It has been usual to take T to be a subset of the real numbers, say $\{1, 2, 3 \cdots\}$ or $[0, \infty)$. However, we need more general index sets such as pairs of integers (labeling the plots in a field trial), the plane (labeling topographic heights), and rectangles in the plane (labeling counts of plants). The great distinction between these indices and those representing time is that the latter have an ordering.

The Daniell–Kolmogorov theorem states that to specify a stochastic process all we have to do is to give the joint distributions of any finite subset $\{Z(t_1), \ldots, Z(t_n)\}$ in a consistent way, requiring

$$P\big(Z(t_i) \in A_i, i = 1, \ldots, m, Z(s) \in \mathbb{R}\big) = P\big(Z(t_i) \in A_i, i = 1, \ldots, m\big)$$

Such a specification is called the *distribution* of the process. We avoid subtle mathematics by only considering a finite number of observations on a stochastic process (except for the differentiability properties in Section 4.4).

We say that the stochastic process is stationary under translations or *homogeneous* if the distribution is unchanged when the origin of the index set is translated. For this to make sense the index set has to be unbounded; it has to be either all pairs of integers or the whole plane. If T

9

is the whole of the plane or three-dimensional space, we can also consider processes that are stationary under rotations about the origin, called *isotropic*.   Homogeneous and isotropic processes are stationary under rigid motions.   The philosophy behind these definitions is discussed in Chapter 1.   Note that they can, at most, be partially checked by, for example, splitting the study region into disjoint parts and checking their similarity.

2.2   COVARIANCES AND SPECTRA

The covariance C and correlation R between $Z(s)$ and $Z(t)$ for two points in T are defined by

$$C(s,t) = E\big[\{Z(s) - E(Z(s))\}\{Z(t) - E(Z(t))\} \big]$$

$$R(s,t) = C(s,t) / \sqrt{\{C(s,s)C(t,t)\}}$$

Homogeneity implies that C and R depend only on the vector $\mathbf{h}$ from s to t, whereas with isotropy they depend only on $d(s,t)$. We will use the notation $C(\mathbf{h})$ or $C(r)$ for these reductions. Note that by symmetry $C(\mathbf{h}) = C(-\mathbf{h})$, but $C((-h_1, h_2))$ may differ from $C((h_1, h_2))$. We will usually plot C in the right half-plane; the other half-plane is found by a half-turn rotation.

In general the distribution of a stochastic process is not completely determined by the mean $m(s) = E[Z(s)]$ and covariance $C(s,t)$. This *is* the case for an important class of processes, the *Gaussian processes* defined by the property that all finite collections $\{Z(t_1), \ldots, Z(t_n)\}$ are joint Normal (that is, every linear combination has a Normal distribution). It is important to know which covariance functions can occur, for given m and C we can construct a Gaussian process via the Daniell–Kolmogorov theorem with that mean and covariance. The necessary and sufficient condition is that C should be *nonnegative definite* and symmetric, that is, that $C(t,s) = C(s,t)$ and

$$\left[\sum_i \alpha_i Z(t_i) \right] = \sum_i \sum_j \alpha_i \alpha_j C(t_i, t_j) \geqslant 0 \qquad (2.1)$$

for all n, $\alpha_1, \ldots, \alpha_n$, $t_1, \ldots, t_n$ (Breiman, 1968, Chapter 11). We often ask that C be strictly positive definite when (2.1) must be nonzero unless all α_i are zero.

This condition of nonnegative definiteness occurs elsewhere, thus enabling us to give examples of valid covariance functions. The characteristic function of a d-dimensional symmetric random vector $\mathbf{X}$ is a nonnegative definite continuous symmetric function on $\mathbb{R}^d$. A continuous homogeneous covariance function of a stochastic process on $\mathbb{R}^d$ will be proportional to such a characteristic function. If $\mathbf{X}$ is rotationally symmetric, the covariance is isotropic. Taking the d-dimensional Cauchy and Normal distributions (with densities proportional to $1/(1+\alpha\|\mathbf{x}\|^2)$ and $\exp-\alpha\|\mathbf{x}\|^2$) shows that e^{-ar} and e^{-ar^2} are both isotropic covariances in any number of dimensions. The *spectral density* f is defined by

$$f(\omega) = \frac{1}{(2\pi)^d} \int \exp\{-i\omega^T\mathbf{h}\} C(\mathbf{h}) \, d\mathbf{h} \tag{2.2}$$

when this integral exists. Then

$$C(\mathbf{h}) = \int \exp\{+i\omega^T\mathbf{h}\} f(\omega) \, d\omega \tag{2.3}$$

Thus $f/C(\mathbf{0})$ is the *pdf* of a random vector with characteristic function $C/C(\mathbf{0})$. For processes on a lattice (2.2) is replaced by a sum and only frequencies for which each component is in the range $[-\pi, \pi]$ are considered, so the integration in (2.3) is restricted to $[-\pi, \pi]^d$. Any nonnegative function that gives a finite value of $C(\mathbf{0})$ in (2.3) is a spectral density.

The spectral density inherits the symmetry condition $f(-\omega)=f(\omega)$ from the covariance function. If the covariance is isotropic $f(\omega)$ becomes a function of $\tau=\|\omega\|$ only, and we have

$$f(\tau) = \frac{1}{2\pi} \int_0^\infty J_0(r\tau) C(r) r \, dr$$

$$C(r) = 2\pi \int_0^\infty J_0(r\tau) f(\tau) r \, d\tau$$

in $\mathbb{R}^2$, where J_0 is the Bessel function (Quenouille, 1949).

The requirement of isotropy on a covariance function is quite restrictive. Schoenberg (1938) conjectured that such a function was continuous, except possibly at the origin. Furthermore, Matérn (1960, pp. 13–19) shows that

$$C(r) \geqslant \inf_u \{k!(2/u)^k J_k(u)\} C(0) \qquad k=(d-2)/2$$

so that isotropic correlations are bounded below by -0.403 in $\mathbb{R}^2$ and

-0.218 in $\mathbb{R}^3$. Isotropic correlation functions are usually specified either by giving an isotropic spectral density or by "mixing" the family e^{-ar}. Suppose we choose a from some distribution, then use a process with correlation function e^{-ar}. The correlation function of the mixed process is $E(e^{-ar})$. This argument shows that any Laplace transform can be a covariance function. (This is the class of functions with $(-1)^n C^{(n)}(r) \geqslant 0$ for $n = 0, 1, 2, \ldots$, and all $r > 0$). One such family of functions are those proportional to $r^\nu K_\nu(ar)$ for $\nu > 0$, with spectral densities proportional to $1/(b + \|\omega\|^2)^{\nu + d/2}$. Here K_ν is a Bessel function. The exponential correlation function is the special case $\nu = \frac{1}{2}$ (Whittle, 1954, 1956, 1963a).

The class of known examples of isotropic correlation functions is not totally satisfactory, for in practice one often finds

$$\mathrm{var}\left\{\int_A Z(\mathbf{x})\,d\mathbf{x}\right\} \sim \mathrm{const}\{\mathrm{area}(A)\}^{(2-\lambda/d)} \tag{2.4}$$

A famous example is given by Fairfield Smith (1938), who found $\lambda \approx 3/2$ for yields from wheat trials. Whittle (1956) showed that (2.4) is equivalent to $C(r)$ and $f(\tau)$ behaving as $r^{-\lambda}$ and $\tau^{(\lambda - d)}$ for large r and small τ, whereas all the standard examples of isotropic covariances decay exponentially at large distances.

One way to form an isotropic process in $\mathbb{R}^d$ is to take a homogeneous process Z_1 with covariance function C_1 on $\mathbb{R}$, to let $Z(\mathbf{x}) = Z_1(x_1)$ and then give the whole of each realization an independent uniformly distributed rotation about the origin in $\mathbb{R}^d$. Then the covariance function of Z is

$$C(r) = \frac{2\Gamma(d/2)}{\sqrt{\pi}\,\Gamma\{\frac{1}{2}(d-1)\}} \int_0^1 C_1(vr)(1 - v^2)^{(d-3)/2}\,dv \tag{2.5}$$

[Matheron, 1973, equation (4.1)]. For $d = 3$ we have the simple results

$$C(r) = \int_0^1 C_1(vr)\,dv, \qquad C_1(r) = \frac{d}{dr}[rC(r)] \tag{2.6}$$

In fact (2.5) is the general form of an isotropic covariance in $\mathbb{R}^d$. It can be re-expressed as

$$C(r) = E\{C_1(rV)\} \tag{2.7}$$

where V is the first coordinate of an independent uniformly distributed point on the surface of the unit ball in $\mathbb{R}^d$. We have noted that $C/C(0)$ is the characteristic function of a random vector $\mathbf{X}$. Because C is isotropic

$\mathbf{X}$ has a rotationally symmetric distribution and

$$C(r) = C(0) E(\exp\{i \|\mathbf{X}\| V\}) \tag{2.8}$$

Comparison of (2.7) and (2.8) shows that we can take $C_1(t) = C(0) E(\exp\{it \|\mathbf{X}\|\})$, which is nonnegative definite and symmetric and so a covariance function in $\mathbb{R}^1$.

The inversion of (2.5) to find C_1 from C provides a way to simulate these processes, as discussed in Section 2.5.

Whittle (1954, 1956), Heine (1955), and Bartlett (1975) discuss the definition of continuous stochastic processes via differential equations. Stochastic differential equations theory is needed to justify their manipulations, which lead to explanatory models for some of the covariances studied here.

2.3   POISSON AND POINT PROCESSES

Point patterns convey a different sort of spatial information from those processes considered so far.   They can be included by defining $Z(\mathbf{x}) = 1$ if there is a point at $\mathbf{x}, 0$ otherwise.   This representation is useless, however, for $P(Z(\mathbf{x}) = 1)$ is usually zero and the distribution of the process then contains no information at all.   We overcome this problem by indexing the stochastic process not by points but by sets, so $Z(A)$ is the number of points in set A.   Every realization of a point process is then a countable set of points, of which a finite number fall within any bounded set.   The points can certainly be located by knowing the counts in each rectangle. In fact, it is sufficient to know which rectangles are nonempty.

The basic point process is a Poisson process, defined by either or both of the following properties:

1.   The number of points in any set A has a Poisson distribution mean $\Lambda(A)$.

2.   Counts in disjoint sets are independent.

Here Λ is a measure giving finite mass to bounded sets, called the *mean measure*.   It is often defined by $\Lambda(A) = \int_A \lambda(\mathbf{x}) d\mathbf{x}$ for some nonnegative bounded function $\lambda(\mathbf{x})$.   A homogeneous Poisson process has mean measure $\Lambda(A) = \lambda$ area (A), where λ is a constant known as the *intensity*, the expected number of points per unit area.   Note that a homogeneous Poisson process is automatically isotropic.

The Poisson process can also be defined in a more constructive way. Consider the process on a bounded set E.   By property 2 it is sufficient to construct the process independently on each of a partition of bounded sets. A binomial process on E is defined by taking N independent identically distributed points in E. To form a Poisson process we take N to have a Poisson distribution with mean $\Lambda(E)$ and then take a binomial process with N points on E; each point having probability density function $\lambda(\mathbf{x})/\Lambda(E)$.

Poisson processes are very convenient building blocks from which to generate other point processes, some examples of which are given in Chapters 6 and 8. They can be regarded as the analogue of independent observations and are often called "random" outside mathematics. (Mathematicians call defining property 2 "purely random" or "completely random.")

The formal mathematics of point processes is given in Kallenberg (1976) and Matthes et al. (1978). The volume edited by Lewis (1972) contains some elementary introductions and almost exclusively one-dimensional applications.

2.4 GIBBS PROCESSES

Gibbs processes are borrowed from statistical physics and provide a way to build new processes from old. We start with a base process P_0 and then define a new process P by giving the probability density ϕ of P with respect to P_0. This density (formally the Radon–Nikodym derivative dP/dP_0) measures the amount by which a particular realization is "more likely" under the new distribution. We can, for instance, prohibit a whole class of realizations by setting $\phi = 0$ on this class.

Unfortunately, ϕ is usually known only up to a normalizing factor that must be chosen to make the new distribution a genuine probability with total mass one. For example, we can define a point process in which no pair of points is closer than D (a model for the centers of nonoverlapping disks of diameter D) by excluding all realizations of a Poisson process in which two points are closer than D. Then, ϕ for the remaining realizations is the reciprocal of the probability that overlap does not occur, and this probability is impossible to compute analytically.

The second problem illustrated by this example is that usually we must confine attention to a bounded region, since in the whole plane the probability of no overlap is zero. Observations are always on a bounded region, so this is not too serious a problem for us. Such processes cannot be truly homogeneous and isotropic, but we can usually arrange for ϕ to be independent of the origin and orientation of the coordinate axes. In statistical physics it is important to be able to define Gibbs processes on the whole of $\mathbb{R}^3$. Two equivalent procedures are (1) to define the process on a bounded region (with suitable boundary conditions) and let this region expand, and (2) to consider the conditional distribution in each bounded region, conditioning on the points in the rest of space. The problem and interest in these approaches is whether they define a unique process. Preston (1974, 1976) and Ruelle (1969, 1970) consider these ideas in great depth.

Markov processes have proved extremely useful in one dimension and several attempts have been made to define analogues on arbitrary spaces including the plane. They all seem less successful because the order property of the line (which means that a single point partitions time into past and future) is fundamental to the theory of Markov processes. Instead of past and future we need a rule that tells us for each pair of points whether or not they are *neighbors*. For observations on a lattice such rules are usually obvious. For points in a plane it is usual to define a pair to be neighbors if they are closer than a given distance. Define the environment $E(A)$ of a set A to be the set of neighbors of points in A. We call a process Markov if the conditional distribution on A given the rest of the process depends only on the process in $E(A)\backslash A$. (This gives the conventional definition for one-dimensional processes in discrete time but a different set of processes in continuous time.)

Two important classes of examples are Gibbs processes with base process independent observations on sites of a lattice, and those with base process a Poisson process on a bounded region. In both cases the base process is Markov, and the Gibbs process is Markov if and only if

$$\phi(\mathbf{x}) = \prod_{\mathbf{y} \subset \mathbf{x}} \psi(\mathbf{y}) \qquad (2.9)$$

and $\psi(\mathbf{y}) = 1$, unless each pair of points in $\mathbf{y}$ is a pair of neighbors. For a point process $\mathbf{x}$ in (2.9) is the collection of points, whereas for a lattice with n sites, $\mathbf{x}$ is the observations at all n sites, and $\mathbf{y} \subset \mathbf{x}$ is a subset of the observations. The lattice case is the so-called Hammersley–Clifford theorem and is much proved. Speed (1978) sketches a neat algebraic treatment. The point process case is due to Ripley and Kelly (1977) whose proof is in the same spirit as Speed's. Equation (2.9) is often re-expressed as

$$\phi(\mathbf{x}) = \exp \sum_{\mathbf{y} \subset \mathbf{x}} V(\mathbf{y})$$

For a system of particles governed by forces $V(\mathbf{y})$ will be the potential of that configuration of particles. For a Markov process $V(\mathbf{y}) = 0$ unless all particles in $\mathbf{y}$ are neighbors. Typically, $V(\mathbf{y}) = 0$ whenever $\mathbf{y}$ contains three or more particles. Then

$$\phi(\mathbf{x}) = A \exp\left[\sum_{\xi \in \mathbf{x}} V(\{\xi\}) + \sum_{\xi, \eta \in \mathbf{x}} V(\{\xi, \eta\}) \right] \qquad (2.10)$$

For an invariant ϕ we take $V(\{\xi\})$ constant and $V(\{\xi, \eta\})$ some function of $d(\xi, \eta)$. Functions ϕ of the form (2.10) then define *pair-potential* processes with

$$\phi(\mathbf{x}) = A b^{\#(\mathbf{x})} \prod_{i < j} h\big(d(x_i, x_j)\big) \qquad (2.11)$$

$\#(\mathbf{x})$ = number of points in realization $\mathbf{x} = \{x_i\}$.

2.5   MONTE CARLO METHODS AND SIMULATION

We have stressed constructive definitions of stochastic processes because we do want to be able to construct realizations, a process known as *simulation*. It is a very intuitive idea that we should be able to assess the fit of a stochastic process to our data by making the computer simulate the process and to compare the outcomes with the data. Suppose we are interested in the distribution of a statistic T, which may be unavailable analytically or have an asymptotic or approximate answer the validity of which is unknown. We can simulate the process m times, calculate the statistic on each simulation, and then compare the empirical distribution with our theoretical work or build up a table of percentage points from the empirical distribution.

Barnard (1963) took this idea a stage further. If the data were actually generated by our hypothetical model, the statistic from the data would be just another simulation, and the probability that it is the rth most extreme is $r/(m + 1)$. By a suitable choice of r and m we can obtain a significance test of any desired size of the null hypothesis that the data were generated by the simulated stochastic process. One-sided or two-sided tests can be achieved by suitably defining "extreme." If the distribution of the test statistic is not continuous, the values of the statistic may tie. In theory, the tied values should be ordered at random. In practice, this will usually be ignored.

The critical region used in Barnard's test procedure is based on quantiles of the empirical distribution of T. Because this is random, we might expect the test to be of lower power than that based on the exact distribution of T were this to be available. The loss in power was investigated by Hope (1968) and Marriott (1979). Their conclusions suggest that $r \geqslant 5$ provides a sensible test with little loss in power.

Monte Carlo methods (when distinguished from simulations) usually involve tricks to increase the "value" of m simulations. Ideas include conditioning on variables and finding the conditional distribution analytically, intentionally using positively or negatively correlated simulations, and simulating some related processes. A problem with using simulation to find empirical distributions is that one is usually interested in the tails of the distribution and most of the simulations provide very little relevant information. One possible application of this set of ideas would be to simulate point processes conditional on the total number N of points and to choose N from a distribution other than its true one to make extreme values of the test statistic more probable. Most of the knowledge about Monte Carlo methods is widely dispersed or even unpublished and uses in spatial statistics are almost nonexistent. Hammersley and Handscomb (1964) give some of the earlier ideas.

Of course, we do have to find efficient ways to simulate our processes. There is a general technique for Gaussian processes. Suppose $\zeta_1, \ldots, \zeta_n$ are independent $N(0,1)$ random variables. Then the covariance matrix of $P\zeta$ is PP^T. Thus to generate joint Normal random variables $Z_1, \ldots, Z_n$ with means $m_1, \ldots, m_n$ and covariance matrix Σ we take a matrix P with $PP^T = \Sigma$ and let $Z_i = m_i + \Sigma p_{ij}\zeta_j$. The simplest way to form P numerically is to note the Cholesky decomposition, which states that there is a lower triangular matrix L with $LL^T = \Sigma$. [Chambers (1977) is a good reference for such numerical methods and for sources of subroutines.] Of course, if a matrix P can be found analytically this should be used. If Σ^{-1} is known, the Cholesky method is useful, for inversion of L is trivial numerically.

Equations (2.5) and (2.6) form the basis of what Matheron (1973) called the "turning band" method to simulate homogeneous isotropic Gaussian processes in $\mathbb{R}^2$ or $\mathbb{R}^3$. Simulate a process Z_1 on $\mathbb{R}$ with covariance C_1 and form Z as described above equation (2.5). Realizations of Z are rather strange (and not joint Normal); hence it is advisable to take m independent copies of Z and form their sum divided by $\sqrt{m}$. This process has the required covariance C. A modification is to take eight independent copies of Z_1 and form

$$Z(\mathbf{x}) = \sum_1^8 Z_i\big[(V_i \mathbf{x})_1\big]$$

where Z_i are independent processes with covariances $C_1/8$ and V_j is a rotation of $2\pi j/8$. This modification makes it particularly easy to simulate the values of Z on a finely spaced lattice in the plane.

A Poisson process is most easily simulated via its constructive definition. Efficient algorithms for sampling Poisson random variables are discussed by Atkinson (1979a, b). Rejection sampling provides a general scheme whereby the points of a binomial process can be found. Suppose we have a supply $Y_1, Y_2, \ldots$, of random variables with probability density g. To generate samples with density h, for each Y_i generate an independent random variable U_i on $[0, 1]$, and accept Y_i if $MU_i \leqslant h(Y_i)/g(Y_i)$; otherwise, try Y_{i+1}. Here M is an upper bound for h/g. For example, to find uniformly distributed points in some irregularly shaped region D, enclose D in $[a, b] \times [c, d]$ and generate points as $(a+(b-a)U_1, c+(d-c)U_2)$, where U_1, U_2 are independent uniform $(0, 1)$ random variables, and accept those points that fall within D.

Rejection sampling can also be applied to whole realizations and so can be used to simulate Gibbs processes with a bounded ϕ. Note that since M need only be an upper bound for ϕ, we do not need to know the awkward normalizing constant A of (2.18). Unfortunately, ϕ is almost always too variable for rejection sampling to be a feasible method and tricks have to be used. Ripley (1979a) gives one trick for Gibbs point processes. Hastings (1970) and Peskun (1973) discuss a family of alternatives to rejection sampling well adapted to Gibbs processes.

CHAPTER 3

Spatial Sampling

This chapter deals with one specific sampling problem. There is a continuous surface $Z(\mathbf{x})$ for which we need the mean value within a region A; we are at liberty to choose n points anywhere within A, and then measure Z at those points. Other sampling problems are discussed in Chapters 6, 7, and 9. Let a be the area (in $\mathbb{R}^2$) or volume (in $\mathbb{R}^3$) of A. We will use

$$\bar{Z} = \frac{1}{n} \sum_1^n Z(\mathbf{x}_i) \qquad \text{to estimate } \tilde{Z}(A) = \int_A Z(\mathbf{x}) \, d\mathbf{x}/a \qquad (3.1)$$

Assuming a continuous surface ensures that the integral in (3.1) makes sense. We assume that Z is a realization of a homogeneous stochastic process with covariance function C and spectral density f. Thus we are taking a superpopulation view of sampling and will be taking expectations both over any randomization involved in sampling and over the surface Z.

3.1 SAMPLING SCHEMES

There are several standard schemes for arranging the n sample points $\{x_1, \ldots, x_n\}$. Figure 3.1 illustrates the possibilities.

Figure 3.1a, c has random elements, whereas Figure 3.1b, d is systematic. *Uniform random sampling*, illustrated in Figure 3.1a operates by choosing each point independently uniformly within A. *Stratified random sampling* takes a uniform random sample of size k from each of m *strata* or subareas, so $n = km$. Figure 3.1c illustrates the usual case of square strata. (Although by no means necessary, square strata are almost universal.) Systematic samples can be of many types. Figure 3.1b illustrates the usual case of an aligned *centric systematic sample*. The only randomization that could be applied to such a sample would be to drop the "centric"

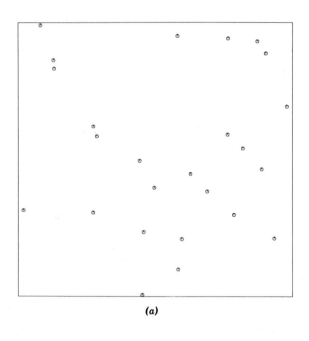

(a)

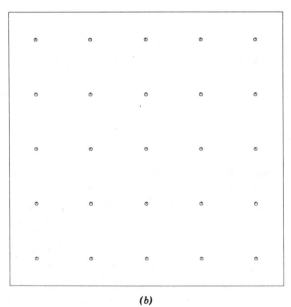

(b)

Fig. 3.1 Four spatial sampling schemes for 25 sample points. (*a*) Uniform random. (*b*) Centric systematic. (*c*) Stratified random. (*d*) Nonaligned systematic.

20

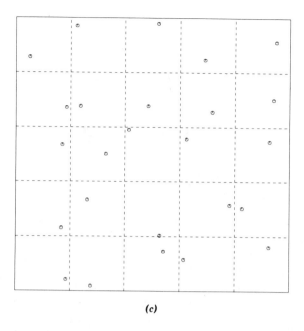

(c)

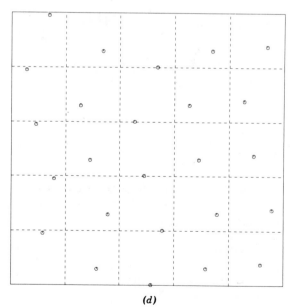

(d)

Fig. 3.1 (*continued*)

and to randomize over the starting position. Figure 3.1d illustrates another type of random sample. The position of the (i, j)th sample is $([(i-1)+\alpha_j]\Delta,[(j-1)+\beta_i]\Delta)$, where Δ is the spacing and (α_i) and (β_j) are prechosen constants between zero and one. To evaluate these sampling schemes we will calculate $n\,\text{var}[\overline{Z}-\tilde{Z}(A)]$.

3.2   ERROR VARIANCES

Let $\mu = E(Z(\mathbf{x}))$ and $\sigma^2 = \text{var}[Z(\mathbf{x})]$

Uniform Random Sampling

We take expectations first over the random positioning of the points $x_1,\dots,x_n$.

$$E\{\overline{Z}-\tilde{Z}(A)|Z\} = \frac{1}{n}\sum_1^n \frac{1}{a}\int_A Z(\mathbf{x})\,d\mathbf{x} - \frac{1}{a}\int_A Z(\mathbf{x})\,d\mathbf{x} = 0$$

$$\text{var}\{\overline{Z}-\tilde{Z}(A)|Z\} = \frac{1}{n^2}\sum \text{var}\{Z(\mathbf{x}_i)\} \qquad \text{by independence}$$

$$= \frac{1}{n}\text{var}[Z(\mathbf{x}_1)] = \frac{1}{na}\int_A [Z(\mathbf{x})-\tilde{Z}(A)]^2\,d\mathbf{x}$$

$$= \frac{1}{na}\int_A [Z(\mathbf{x})-\mu]^2\,d\mathbf{x}$$

$$- \frac{1}{na^2}\int_A\int_A [Z(\mathbf{x})-\mu][Z(\mathbf{y})-\mu]\,d\mathbf{x}\,d\mathbf{y}$$

$$\text{var}\{\overline{Z}-\tilde{Z}(A)\} = E\left[\text{var}\{\overline{Z}-\tilde{Z}(A)|Z\}\right] = \frac{1}{n}\left[\sigma^2 - E\{C(\mathbf{X},\mathbf{Y})\}\right] \quad (3.2)$$

where $\mathbf{X},\mathbf{Y}$ are independent uniformly distributed points in A.

For A large compared with the effective range of the covariance function the second term will be negligible.

Stratified Random Sampling

Suppose A is partitioned into m strata $S_1,\dots,S_m$ each of area s. First averaging over the random sampling

$$E\{\overline{Z}|Z\} = E\left(\frac{1}{m}\sum\overline{Z}_i \,\middle|\, Z\right) = \frac{1}{m}\sum\tilde{Z}(S_i) = \tilde{Z}(A)$$

$$\text{var}\{\overline{Z}-\tilde{Z}(A)|Z\} = \frac{1}{m^2}\sum_i \frac{1}{ks}\left[\int_{S_i}(Z(\mathbf{x})-\mu)^2\,d\mathbf{x}\right.$$
$$\left. - \int_{S_i}\int_{S_i}(Z(\mathbf{x})-\mu)(Z(\mathbf{y})-\mu)\,d\mathbf{x}\,d\mathbf{y}\right]$$

using the results above for the averages $\overline{Z}_1,\ldots,\overline{Z}_m$ for each stratum. Taking expectations over the process

$$\text{var}(\overline{Z}-\tilde{Z}(A)) = E\left[\text{var}\{\overline{Z}-\tilde{Z}(A)|Z\}\right]$$
$$= \frac{1}{m^2k}\sum\left[\sigma^2 - E\{C(\mathbf{X}_i,\mathbf{Y}_i)\}\right]$$
$$= \frac{1}{n}\left[\sigma^2 - \overline{E\{C(\mathbf{X}_i,\mathbf{Y}_i)\}}\right] \tag{3.3}$$

where now $\mathbf{X}_i,\mathbf{Y}_i$ are random points in the ith stratum. The point of stratified sampling is that we should choose strata sufficiently small to make $E\{C(\mathbf{X},\mathbf{Y})\}$ as large as possible, so that (3.3) is less than (3.2).

Systematic Sampling

Let the sample be at points labeled by $\mathbf{u}\in A$. Then

$$E(\overline{Z}) = \frac{1}{n}\sum_{\mathbf{u}} E(Z(\mathbf{u})) = \frac{1}{n}\sum_{\mathbf{u}}\mu = \mu$$

$$\text{var}(\overline{Z}-\tilde{Z}(A)) = \frac{1}{n^2}E\left[\sum_{\mathbf{u}}\{Z(\mathbf{u})-\tilde{Z}(A)\}\right]^2$$
$$= \frac{1}{n^2}\sum_{\mathbf{u},\mathbf{v}}C(\mathbf{u},\mathbf{v}) - \frac{2}{n}\sum_{\mathbf{u}}E\{(Z(\mathbf{u})-\mu)(\tilde{Z}(A)-\mu)\}$$
$$+ E\{\tilde{Z}(A)-\mu\}^2$$
$$= \frac{1}{n^2}\sum_{\mathbf{u},\mathbf{v}}C(\mathbf{u},\mathbf{v}) - 2\sum_{\mathbf{u}}\frac{1}{an}\int_A C(\mathbf{u},\mathbf{y})\,d\mathbf{y}$$
$$+ \frac{1}{a^2}\int_A\int_A C(\mathbf{x},\mathbf{y})\,d\mathbf{x}\,d\mathbf{y} \tag{3.4}$$

To make further progress we assume there is a $(r\times s)$ grid of points, of step size Δ in both directions, and use stationarity with $C(\mathbf{x},\mathbf{y})=c\{\mathbf{x}-\mathbf{y}\}$ to find

$$n\,\mathrm{var}\big(\bar{Z}-\tilde{Z}(A)\big)\approx\sum_{-r}^{r}\sum_{-s}^{s}\left(1-\frac{|u|}{r}\right)\left(1-\frac{|v|}{s}\right)c\{\Delta(u,v)\}$$

$$-\frac{1}{\Delta^2}\int_{-r\Delta}^{r\Delta}\int_{-s\Delta}^{s\Delta}\left(1-\frac{|x|}{r\Delta}\right)\left(1-\frac{|y|}{s\Delta}\right)c\{(x,y)\}\,dx\,dy$$

the approximation involving neglecting the difference between r and $r+1$, s and $s+1$. Again approximating by assuming that the range of c is small compared with $r\Delta$ and $s\Delta$ we find

$$n\,\mathrm{var}\big\{\bar{Z}-\bar{Z}(A)\big\}\approx\sum_{u,\,v\text{ integers}}c\{\Delta(u,v)\}$$

$$-\frac{1}{\Delta^2}\int_{-\infty}^{\infty}\int_{-\infty}^{\infty}c\{(x,y)\}\,dx\,dy \qquad (3.5)$$

which in terms of the spectral density f is

$$n\,\mathrm{var}\big\{\bar{Z}-\bar{Z}(A)\big\}=\frac{4\pi^2}{\Delta^2}\left[\sum_{\substack{\mu,\,\nu\\ \text{integers}}}f\left\{\left(\frac{2\pi\mu}{\Delta},\frac{2\pi\nu}{\Delta}\right)\right\}-f(0,0)\right] \qquad (3.6)$$

Several conclusions can be drawn from (3.2), (3.3), and (3.6). For an isotropic covariance $C(\mathbf{X},\mathbf{Y})=c(\|\mathbf{X}-\mathbf{Y}\|)$ so

$$E(C(\mathbf{X},\mathbf{Y}))=\int_0^{\infty}c(r)b_A(r)\,dr$$

where $b_A(r)$ is the density function of the distribution of the distance between two points uniformly distributed in A. Clearly, this is largest for compact regions such as circles or squares. This is relevant to the choice of stratum shape. If we assume all strata are congruent to S, (3.3) becomes

$$n\,\mathrm{var}\big(\bar{Z}-\tilde{Z}(A)\big)=\sigma^2\left[1-\int_0^{\infty}R(r)b_S(r)\,dr\right] \qquad (3.7)$$

The gain from stratification will be most when $R(r)$ is large for all r up to the diameter of S, but becomes negligible on the scale of S. This suggests that for monotonically decreasing correlation functions we should take small strata; hence k will be small.   For a square stratum of side Δ (3.3) becomes

$$\sigma^2 - \frac{1}{\Delta^2} \int_{-\Delta}^{\Delta}\int_{-\Delta}^{\Delta} c((x,y))\left(1-\frac{|x|}{\Delta}\right)\left(1-\frac{|y|}{\Delta}\right) dx\, dy$$

$$= \int f(\omega)\, d\omega - \int f((\mu,\nu))\left\{ \int_{-\Delta}^{\Delta}\left(1-\frac{|x|}{\Delta}\right) e^{-i\mu x}\frac{dx}{\Delta}\right.$$

$$\left. \times \int_{-\Delta}^{\Delta}\left(1-\frac{|y|}{\Delta}\right) e^{-i\nu y}\frac{dy}{\Delta}\right\} d\omega$$

$$= \int f((\mu,\nu))\left[1-\frac{4}{\Delta^2\mu^2\nu^2}(1-\cos\mu\Delta)(1-\cos\nu\Delta)\right] d\mu\, d\nu \qquad (3.8)$$

Compare (3.8) with (3.6).   For frequencies at which $(\mu\Delta)$ and $(\nu\Delta)$ are both small, the second factor in (3.8) is near 1, so both formulas are approximations to $\int f(\omega)\, d\omega$ except that both reduce the weights given to low frequencies, more sharply for (3.6) than for (3.8).   Thus, if low frequencies are dominant (corresponding to strong local positive correlation), both stratified random and systematic sampling should do well relative to uniform random sampling.   Furthermore, we would expect systematic sampling to be best unless there is a sharp peak in the spectral density at one of the frequencies summed in (3.6).   In practice, this would only occur if the process has a strong periodicity with wavelength the basic sampling interval Δ along either axis or with wavelength $\sqrt{2}\,\Delta$ along a diagonal.

3.3   ESTIMATING SAMPLING ERRORS

For uniform random sampling we would compute the sample variance

$$s^2 = \sum_{1}^{n}\left[Z(\mathbf{x}_i)-\bar{Z}\right]^2/(n-1) \qquad (3.9)$$

Taking expectations over the randomization we get

$$a^{-1}\int_A\left[Z(x)-\tilde{Z}(A)\right]^2 dx$$

which gives us

$$\sigma^2 - E(C(\mathbf{X}, \mathbf{Y}))$$

when we average over the process. From (3.2) s^2/n is an unbiased estimator of the sampling error variance.

For stratified random sampling with k, the number of points per stratum, at least 2 we can apply (3.9) to each stratum. The average within stratum variance is then an unbiased estimator of $n\,\mathrm{var}(\bar{Z} - \tilde{Z}(A))$. However, we have seen that the latter is smallest when the strata are chosen as small and compact as possible. Ideally, we would choose n strata with $k = 1$. In this case, or with systematic sampling, we have no information left with which to estimate the sampling error.

Clearly, with systematic sampling we will never be able to assess the variability due to the positioning of the grid, so we must assume that this is small; that is, that there is no periodicity in the process at the sampling wavelength. Finney (1948, 1950, 1953) has warned of this problem and presented an example of apparent periodicity in a forest survey. Matérn (1960) pointed out that this could be pseudo-periodicity as described in time series by Yule. However, (3.6) makes clear that either true or pseudo-periodicity is very detrimental to the precision of a systematic

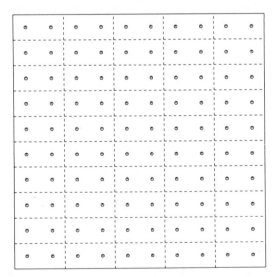

Fig. 3.2 Artificial strata for the estimation of the sampling variance with a 10×10 centric systematic sample.

sample. Milne (1959) found inconsistencies in Finney's example and suggested that the periodicity is, in fact, caused by defects in the sampling, for the forest was assigned to the group of enumerators in a periodic manner. Even if there were an unsuspected periodicity in the process being sampled, Milne concludes that "the danger to centric systematic sampling from unsuspected periodic variation is so small as to be scarcely worth a thought."

There remains the problem of how to estimate the sampling variance in systematic samples or with stratified random sampling with one sample per stratum. Two possibilities are to use formula (3.9) and regard s^2/n as the error variance or to impose larger strata on the sample, as illustrated in Figure 3.2, and to use the stratified sampling formula formed by averaging (3.9) over each of these artificial strata and dividing by n. Milne has an empirical study that suggests that either method gives a good idea of the true sampling variance. Matérn (1960) shows theoretically that this procedure may (slightly) overestimate the sampling variance.

3.4 OPTIMAL LOCATION OF SAMPLES

Dalenius et al. (1961) consider the optimal location of sampling points for Zubrzycki's process (Section 4.4). They show that a sampling scheme should give as high a degree of overlap as possible between disks centered on the sample points with radius R, that defining the process. Clearly, this needs a systematic sample. They show that the optimal schemes (in the sense of minimum achievable sampling variance per sample point) are various triangular, rectangular, and hexagonal lattices, the choice depending on the number of samples per unit area and R. They conjecture that an equilateral triangular lattice is optimal for all exponential covariance functions and hence (by mixing) for all completely monotonic isotropic covariances.

References

Arvantis and O'Regan (1972), Barrett (1965), Dalenius et al. (1961), Finney (1948, 1950, 1953), Hannan (1962), Hasel (1938), Holmes (1970), Martin (1979), Matérn (1960), Milne (1959), Osbourne (1942), Payendeh (1970a, b), Quenouille (1949).

CHAPTER 4

Smoothing and Interpolation

Throughout this chapter we assume that we are given observations $Z(x_i)$, and we wish to map the surface Z within a region D. The sample points $x_1, \ldots, x_N$ are usually, but not always, inside D. They might be on a regular grid, as suggested by the theory of Chapter 3, or they might be those points at which data is available, chosen for other reasons. For instance, networks of rain gauges are set up where observers are available, and data on oil and mineral fields are available where drilling occurred (at spots thought to be fruitful) and from those parts of the field that are being exploited. Under these circumstances, the sample mean may be seriously biased as an estimator of the mean level of the surface within D. An alternative is to fit an interpolating surface to the data values and to find its average. Such a surface can have independent uses. In mining, a map of the mineral grade (for example, percentage of copper) will help plan the mining operation as well as give information on which parcels will have a high enough average grade to make processing economic. It may also be helpful to have a smoothed map to indicate the broad features of the data as well as an interpolated map for prediction. Indeed, if the data are themselves from samples as in mining or soil surveys, they may be thought to have an appreciable measurement error, possibly making interpolation undesirable.

Often field workers will have extra, qualitative, information on the surface to be analyzed, or such data may also be available by analogy, for instance, from similar mines that have been worked out. It is important to have some way of introducing this information. Some of the methods to be described will produce only one map from a given set of observations, whereas others have several parameters that may be "fine tuned." These are used most easily at an interactive graphics terminal, so that the visual effect can be gauged.

Surfaces over a two-dimensional domain can be represented on a plotter or in print in one of two ways. Both are illustrated in Figure 1.2. Most users find it easiest to grasp the broad features from perspective plots and fine details from contour maps. With perspective plots it is often important to pick the best view, and even with contour plots it is useful to have available interactively the facility to add lines at critical contour levels and to explore parts of the map in detail. Unfortunately these opportunities are denied to the reader, who has to accept my choices. The contouring algorithm used is discussed in Section 4.5.

The first four sections discuss four conceptual approaches. Trend surfaces are the generalizations to more than one dimension of curve-fitting by least squares. The spatial analogues of spline interpolation are discussed in Section 4.3. The other two methods are extensions of those used to forecast time series: moving averages and the Wiener–Kolmogorov theory (often regarded as the "Box–Jenkins method"). All methods are discussed in terms of two examples and compared at the end of Section 4.4.

4.1 TREND SURFACES

Multidimensional generalizations of polynomial regression go back to "Student" in 1914; however, they were popularized in the earth sciences by Grant (1957) and Krumbein (1956), about which time computers made the computational task less daunting. This is a smoothing technique—the idea is to fit to the data by least squares a function of the form

$$f((x, y)) = \sum_{r+s \leqslant p} a_{rs} x^r y^s \qquad (4.1)$$

of which the first few functions are:

a	flat
$a + bx + cy$	linear
$a + bx + cy + dx^2 + exy + fy^2$	quadratic

These formulas cover two dimensions. They have obvious extensions to three or more dimensions. The integer p is the *order* of the trend surface. There are $P = (p+1)(p+2)/2$ coefficients, which are normally chosen to minimize

$$\sum_{1}^{N} \{ Z(\mathbf{x}_i) - f(\mathbf{x}_i) \}^2 \qquad (4.2)$$

Linear, quadratic, and cubic fits to 10 points are illustrated in Figure 4.1.
Notice that a cubic surface has 10 parameters and so provides an exact fit
to the data.   An undesirable feature of trend surfaces is shown, a tend-
ency to wave the edges to fit points in the center.   This effect is well
known in polynomial regression, but is more severe in two or more
dimensions where there is more boundary to be affected.

We can rewrite (4.1) to make (4.2) a standard least-squares or multiple
regression problem.   If we label $1, x, y, x^2, xy, y^2, \ldots,$ as $f_1(\mathbf{x}), \ldots, f_P(\mathbf{x})$
and the coefficients as $\beta_1, \ldots, \beta_P$, the problem is multiple regression of
$Z(\mathbf{x}_i)$ on $(f_1(\mathbf{x}_i), \ldots, f_P(\mathbf{x}_i))$ and could be given to a statistical package.
Now polynomial regression is well known as an ill-conditioned least-squares
problem that needs careful numerical analysis, and this is equally true of
polynomial trend surfaces.   It is desirable to rescale D to about a square
with sides $[-1, +1]$ to avoid extremely large or small values in $f_i(\mathbf{x})$.
Orthogonal polynomials are often used for polynomial regressions, especially with regularly spaced data.   This approach has been extended to
trend surfaces by Grant (1957) and Whitten (1970, 1972, 1974a).   Modern
multiple regression techniques (for which see Chambers, 1977 or Seber,
1976) rely on orthogonalizing the functions $f_1, \ldots, f_P$ and make the orthogonal–polynomial technique unnecessary.   Suppose we write the surface as

$$Z(\mathbf{x}) = \mathbf{f}(\mathbf{x})^T \boldsymbol{\beta} + \varepsilon(\mathbf{x}), \qquad \mathbf{f}(\mathbf{x}) = \begin{bmatrix} f_1(\mathbf{x}) \\ \vdots \\ f_P(\mathbf{x}) \end{bmatrix} \tag{4.3}$$

and

$$\mathbf{Z}_N = F\boldsymbol{\beta} + \varepsilon, \qquad F = \begin{bmatrix} \mathbf{f}(\mathbf{x}_1)^T \\ \vdots \\ \mathbf{f}(\mathbf{x}_N)^T \end{bmatrix}, \qquad \mathbf{Z}_N = \begin{bmatrix} Z(\mathbf{x}_1) \\ \vdots \\ Z(\mathbf{x}_N) \end{bmatrix}$$

These algorithms find an orthogonal matrix Q $(Q^T Q = QQ^T = I)$ and an
upper triangular matrix $\tilde{R}$ so that

$$QF = R = \begin{bmatrix} \tilde{R} \\ 0 \end{bmatrix} \qquad Q_{N \times N}, F_{N \times P}, \tilde{R}_{P \times P} \tag{4.4}$$

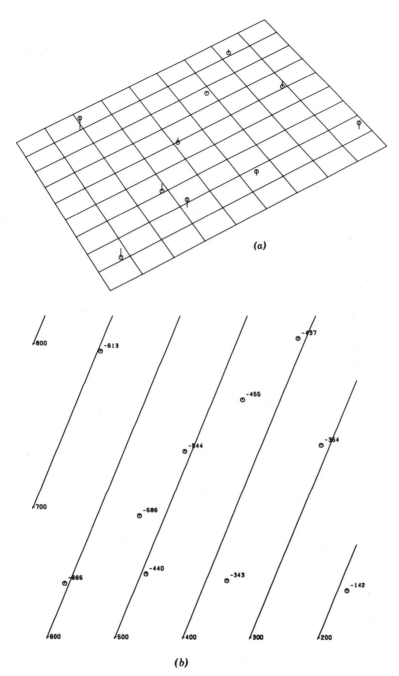

Fig. 4.1 Least-squares trend surfaces for 10 data points (from Davis 1973, p. 328). (a), (b) Linear.

31

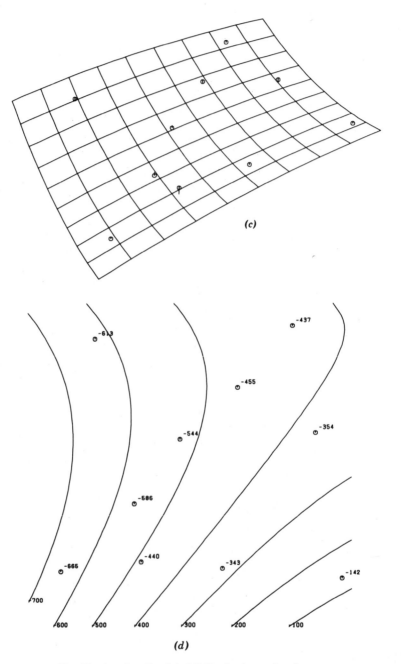

(c)

(d)

Fig. 4.1 (*continued*). (*c*), (*d*) Quadratic trend surface.

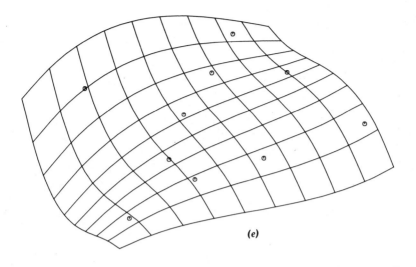

(e)

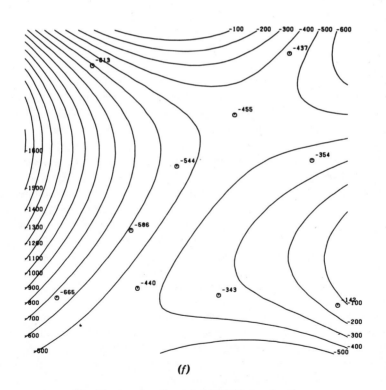

(f)

Fig. 4.1 (*continued*). (*e*), (*f*) Cubic trend surface.

33

Then the least-squares estimate of β is found by

$$\sum_1^N \left(Z(\mathbf{x}_i) - \mathbf{f}(\mathbf{x}_i)^T\beta \right)^2 = (\mathbf{Z}_N - F\beta)^T(\mathbf{Z}_N - F\beta)$$

$$= \left[Q(\mathbf{Z}_N - F\beta) \right]^T \left[Q(\mathbf{Z}_N - F\beta) \right]$$

$$= \left[Q\mathbf{Z}_N - R\beta \right]^T \left[Q\mathbf{Z}_N - R\beta \right]$$

$$= \mathbf{Y}_2^T\mathbf{Y}_2 + (\mathbf{Y}_1 - \tilde{R}\beta)^T(\mathbf{Y}_1 - \tilde{R}\beta) \qquad (4.5)$$

where

$$Q\mathbf{Z}_N = \begin{bmatrix} \mathbf{Y}_1 \\ \mathbf{Y}_2 \end{bmatrix}, \qquad (\mathbf{Y}_1)_{P \times 1}, \qquad (\mathbf{Y}_2)_{(N-P) \times 1}$$

Provided $\tilde{R}$ has full rank, we obviously get a minimum sum of squares when $\mathbf{Y}_1 = \tilde{R}\beta$,

$$\hat{\beta} = \tilde{R}^{-1}\mathbf{Y}_1 \qquad (4.6)$$

Since $\tilde{R}$ is upper triangular, it is easy to solve (4.6) for $(\beta_P, \beta_{P-1}, \ldots, \beta_1)$. The rank of $\tilde{R}$ is that of F, so our condition is that $f_1, \ldots, f_P$ be linearly independent when evaluated at $(\mathbf{x}_1, \ldots, \mathbf{x}_N)$ and principally excludes all data points lying on a line. We will assume that it is satisfied.

The usual statistical model associated with (4.3) assumes that the "errors" $\varepsilon(\mathbf{x})$ are independent Normally distributed random variables with zero mean and variance σ^2 for all $\mathbf{x}$. This is rather unrealistic, for we would expect deviations from a trend surface to be positively correlated over short distances. We will find better assumptions in Section 4.4, but if for the moment we accept this distribution, we find that the covariance matrix of the $\hat{\beta}$'s is $\sigma^2(F^TF)^{-1} = \sigma^2(\tilde{R}^T\tilde{R})^{-1}$. Thus the variance of $\mathbf{f}(\mathbf{x})^T\beta$, the value of the surface at the point $\mathbf{x}$, is

$$\sigma^2 \| \tilde{R}^{-T}\mathbf{f}(\mathbf{x}) \|^2 \qquad (4.7)$$

where $\|x\|^2$ is the squared length of a vector and $\tilde{R}^{-T}$ is shorthand for $(\tilde{R}^T)^{-1} = (\tilde{R}^{-1})^T$. From this model we can perform the usual F tests of multiple regression to see whether to increase or decrease the order of the trend surface. But if the "errors" are positively correlated, they jointly carry less information than the theory suggests and we would usually be led to fit a surface of too high an order.

There has been much discussion on the effect of the pattern of the data points on the fitted trend surface. Davis (1973, pp. 349–352) illustrates some experiments with samples from *polynomial* surfaces that suggest that whereas it is important to have data points throughout D, it does not matter much if they are clustered. Gray (1972) and Robinson (1972) concur. It does seem intuitively obvious that when the fit of the surface is not good, clusters of points are equivalent to weighting that part of domain D near the clusters to have a good fit. An uneven distribution of clusters can distort the fitted surface.

The astute reader may wonder why we chose only the full linear and quadratic surfaces rather than introduce the second-order terms one at a time. The fitted surface should be invariant under rigid motions, since the coordinate scheme chosen is purely a convenience and the fitted surface should not depend on it. Cliff and Kelly (1977) consider invariant trend surfaces and find some invariant combinations of the parameters, but only the full surfaces are invariant. If combinations of terms are allowed there are a few other invariant surfaces, for instance

$$a + bx + cy + d(x^2 + y^2)$$

These exceptional invariant surfaces are all of even order and have the terms of highest degree of the form $a(x^2 + y^2)^r$.

Other families of surfaces have been suggested but are apparently rarely used. The main suggestion is a limited Fourier expression

$$f(x, y) = \sum_{0 < i, j < r} \left[a_{ij} \sin(i\omega_1 x) \sin(j\omega_2 y) + b_{ij} \sin(i\omega_1 x) \cos(j\omega_2 y) \right.$$

$$\left. + c_{ij} \cos(i\omega_1 x) \sin(j\omega_2 y) + d_{ij} \cos(i\omega_1 x) \cos(j\omega_2 y) \right] \qquad (4.8)$$

Such a surface is not invariant and has a large number of coefficients $[4r^2 + 4r + 1$, since $\sin(0) = 0]$, so even with $r = 3$ there are 49 coefficients plus ω_1 and ω_2 to be fitted or chosen. With small values of r the surface shows clear ripples at the fundamental frequencies ω_1 and ω_2 in the x and y directions and so is really only suitable for periodic phenomena. With data on a rectangular grid this model is much more useful. In particular, we shall see in Chapter 5 that there are dramatically better computational procedures for that case.

By itself, trend surface analysis is best at showing broad features of the data. Its main use, however, is either to remove those broad features to allow some other technique to work on the residuals or as part of the computation for the stochastic process prediction approach of Section 4.4.

References and Applications

Agterburg (1968), Berry (1971), Chayes (1970), Chorley and Haggett (1965), Cliff and Kelly (1977), Cole (1970), Fairburn and Robinson (1969), Grant (1957), Gray (1972), Howarth (1967), C.A.M. King (1969), Krumbein (1956, 1959, 1963), Link and Koch (1970), Norcliffe (1969), Olsson (1968), Parsley (1971), Reilly (1969), Rhodda (1970), G. Robinson (1970, 1972), Tarrant (1970), Tobler (1964), Torelli and Tomasi (1977), Trapp and Rockaway (1977), Unwin (1970, 1973), Unwin and Lewin (1971), Watson (1972), Whitten (1970, 1972, 1974a), Wilson (1975).

4.2 MOVING AVERAGES

A very simple way to smooth or interpolate is to predict the value of a surface by a weighted average of the values at the data points. These weights are chosen as a function of the distances:

$$\hat{Z}(\mathbf{x}) = \sum_1^N \lambda_i Z(\mathbf{x}_i) \qquad \Sigma \lambda_i = 1, \lambda_i \propto \omega(d(\mathbf{x}, \mathbf{x}_i)) \qquad (4.9)$$

Clearly, the surface will interpolate if $\omega(d) \to \infty$ as $d \to 0$. Common choices of ω seem to be d^{-r}, $e^{-\alpha d}$, and $e^{-\alpha d^2}$.

If we suppose ω is smooth, clearly $\hat{Z}(\mathbf{x})$ is differentiable, except possibly at a data point. For points at distances from $\mathbf{x}_1$ small compared with the interpoint distances,

$$\hat{Z}(\mathbf{x}) - Z(\mathbf{x}_1) = \left[\sum_2^N \omega_i \{ Z(\mathbf{x}_i) - Z(\mathbf{x}_1) \} \right] \Big/ \left[\omega_1 + \sum_2^N \omega_i \right]$$

where $\omega_i = \omega(d(\mathbf{x}, \mathbf{x}_i))$ and we can regard all terms except ω_1 as constant. Thus Z is differentiable at $\mathbf{x}_i$ if and only if $d/\omega(d) \to 0$ as $d \to 0$. The same argument applies to all other data points. There is also a constraint on $\omega(d)$ for large d. Suppose the region D is expanded. We would expect (for a homogeneous pattern of data points) about const$\cdot d \Delta d$ data points from distance d to $d + \Delta d$ from a point $\mathbf{x} \in D$, and these contribute about const$\cdot d \Delta d \cdot \omega(d)$ to the total of the weights. Thus unless $\int_1^\infty t\omega(t) \, dt$ is finite, the value $\hat{Z}(\mathbf{x})$ is *not* an average of local values, but depends entirely on how large D is chosen to be. We need $\omega(d)$ to decay faster than d^{-2} for large d. The effect of violating either condition is to give a fairly smooth surface with sharp spikes at the data points to ensure interpolation. Figure 4.14 gives an example. Usually $\omega(d)$ is chosen to be zero beyond some arbitrarily chosen radius.

Fig. 4.2 Susceptibility of moving averages to clustered data points. The 10 clustered points will dominate the rest for estimation at ×.

There are further drawbacks. If the data points lie on an inclined plane, $\hat{Z}$ will be nothing like a plane. The method is susceptible to clusters in the data points, which in mining usually occur at "high" points on the surface (see Figure 4.2). One way out is to estimate the value at **x** from the few nearest points in each quadrant centered on **x**. The problem of planar data is worse for a quadratic surface, for $\hat{Z}(\mathbf{x})$ is bounded above and below by the maximum and minimum of the data values and so cannot follow a quadratic peak.

These problems are addressed by the distance-weighted least-squares method of Pelto et al. (1968) and McLain (1974). The idea is to solve the weighted least-squares problem

$$\min \sum_{1}^{N} \omega\big(d(\mathbf{x},\mathbf{x}_i)\big)\big[\, Z(\mathbf{x}_i)-\mathbf{f}(\mathbf{x}_i)^T\boldsymbol{\beta}\,\big]^2 \qquad (4.10)$$

for each point **x**, and let $\hat{Z}(\mathbf{x})$ be the fitted value $\mathbf{f}(\mathbf{x})^T\boldsymbol{\beta}(\mathbf{x})$. Weighted least squares can be performed by a simple modification of the procedure in Section 4.1—form a diagonal matrix W with entries $\sqrt{\omega(d(\mathbf{x},\mathbf{x}_i))}$ and apply these procedures to WZ_N and WF. Distance-weighted least squares is also vulnerable to clusters. We still need conditions on $\omega(d)$. $\omega(d)\to\infty$ for $d\to0$ ensures interpolation, whereas $d^2\omega(d)\to0$ as $d\to\infty$ gives a "local" character as before. In general, the surface is as many times differentiable as $1/\omega(|u|)$ is as a function of $u\in\mathbb{R}$. For $\omega(d)\sim d^{-p}$ for small d the surface will be differentiable up to the integer less than p unless p is even, in which case the surface is infinitely differentiable (the case treated by Pelto et al., 1968). The following argument is a proof for a linear surface in one dimension. It extends to the general case, but becomes notationally complex!

Parametrize the line as $f(\tilde{x})=a+b(\tilde{x}-x)$. Consider x near x_1, say. Then (4.10) becomes

$$\min_{a,b}\left[\sum_{i>1}\omega_i(Z(x_i)-a-b[\,x_i-x\,])^2+\omega_1(Z(x_1)-a)^2\right]$$

Differentiating with respect to a and b and eliminating b we find

$$\hat{a} = \frac{\omega_1 Z(x_1) + \sum_{i>1} \omega_i (1 + C[x - x_i]) Z(x_i)}{\omega_1 + \sum_{i>1} \omega_i + C \sum_{i>1} \omega_i (x - x_i)}$$

where

$$C = \frac{\sum_{i>1} \omega_i (x_i - x)}{\sum_{i>1} \omega_i (x_i - x)^2}$$

Thus since $\hat{Z}(x) = \hat{a}$,

$$\hat{Z}(x) - Z(x_1) - \hat{b}(x - x_1) \sim \frac{1}{\omega_1}\left[\sum_{i>1} \omega_i (1 + C[x - x_i])\{Z(x_i) - Z(x_1)\} \right]$$

$$(4.11)$$

The formula for $\hat{b}$ shows that it is effectively constant as $x \to x_1$, so the differentiability results follow from (4.11). Indeed, the clue to the general problem is to note that as $x \to x_1$, all the coefficients approach the solution of the problem (4.10) with the surface constrained to pass through $Z(x_1)$.

McLain suggested $\omega(d) = (e^{-d^2/\alpha})/(d^2 + \varepsilon)$, where ε is a small constant to avoid numerical overflow, and α is the average squared nearest-neighbor distance.

4.3 TESSELLATIONS AND TRIANGULATIONS

Many one-dimensional interpolation techniques are based on fitting some curve in each of the intervals between the data points and choosing the parameters of the curve to give continuity of a certain number of derivatives at each data point. These are usually called *spline* methods. This section deals with the attempts at defining spatial splines, which are a much harder subject. Whereas the data points in one dimension divide the real line into intervals, it is not obvious how to use our data points in the plane to divide D into regions. Figure 4.3 illustrates the two basic constructs. To each point we associate a Dirichlet cell (also known as a Voronoi or Thiessen polygon), which is that part of D that is nearer to that

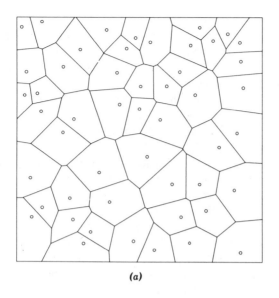

(a)

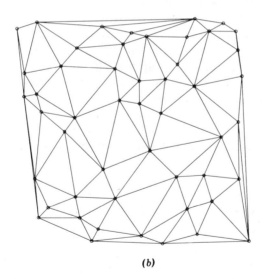

(b)

Fig. 4.3 (*a*) Dirichlet tessellation. (*b*) Delaunay triangulation.

data point than any other.   Green and Sibson (1978) provide an algorithm
to construct these cells, which they refer to as *tiles*.   (Their program was
used in constructing all the Dirichlet tessellations in this book.)   From the
Dirichlet cells we can form the Delaunay triangulation, joining points for
which the associated polygons have an *edge* in common.   These constructs
are well known in packing theory; see C. A. Rogers (1964) and Miles
(1974a) for some properties.   Other triangulations have been proposed.
Sibson (1978) has shown that the Delaunay triangulation uniquely achieves
the Lawson criterion (cf. Lawson, 1977).   Each pair of adjacent triangles
forms a quadrilateral.   Lawson required that the smallest of the six angles
in the two triangles should be larger for this division of a convex quadrilateral than that given by the other diagonal, as explained in Figure 4.4.
An algorithm for this triangulation is contained in Akima (1978).   Another
criterion, attributed by McLain (1976) to Pitteway, is that any point within
any triangle should have as its closest data point one of the vertices of that
triangle.   Although McLain gives an algorithm, there are patterns of data
points for which no triangulation is Pitteway.   Note that there are degen-
erate cases in which the Delaunay–Lawson triangulation is not unique,
and the formal definition given above does not give a triangulation.
Figure 4.5 illustrates such a pattern.

The oldest and simplest methods based on tessellations are the method
of polygons of influence, in which the surface is assigned over each tile the
value at its defining data point, and fitting a plane to each triangle as
described by Bengtsson and Nordbeck (1964).   Both methods produced
unacceptable surfaces but could be computed by hand.   Note that the
Delaunay triangulation is defined only within the convex hull of the data
points and so usually does not cover D.

The method proposed by McLain uses the distance-weighted least-
squares technique.   He calculated a surface $f_1(x)$, $f_2(x)$, $f_3(x)$ at each

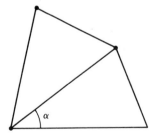
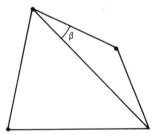

Fig. 4.4 The Lawson criterion. The left-hand triangulation is chosen because angle α is
larger than angle β.

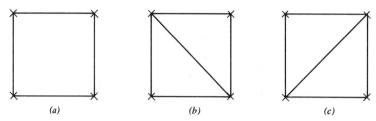

Fig. 4.5 Degeneracy in the Delaunay triangulation. The formal definition gives (a). Both (b) and (c) are valid triangulations.

vertex of a triangle by distance-weighted least squares, and then used

$$f(\mathbf{x}) = \frac{w_1(\mathbf{x})f_1(\mathbf{x}) + w_2(\mathbf{x})f_2(\mathbf{x}) + w_3 f_3(\mathbf{x})}{w_1(\mathbf{x}) + w_2(\mathbf{x}) + w_3(\mathbf{x})} \qquad (4.12)$$

within that triangle. (He used quadratic surfaces $f_i(\mathbf{x})$, but that is not essential.)   For the notation needed, see Figure 4.6.   For continuity of the surface, $w_i(\mathbf{x})$ vanishes on the edge opposite vertex i.   McLain claims in his correction note that the fitted surface has continuous derivatives of order up to and including $(n-1)$ if

$$w_1(\mathbf{x}) = d_1^n (d_2^n d_{23}^n + d_3^n d_{32}^n) \qquad (4.13)$$

and w_2 and w_3 are found by permuting the vertex labels.   Whether this surface interpolates depends on the weighting used to find the distance-weighted least-squares surface at the vertex.   As before, we need to give infinite weight to the vertex to ensure interpolation.

Sibson (1980b, c) has described a method he calls "natural neighbourhood interpolation," which is closely related to McLain's.   For data points $\mathbf{x}_i$ and $\mathbf{x}_n$ define $\lambda_n(\mathbf{x}_i)$ to be the proportion of the total area of the tile of

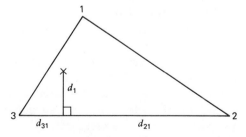

Fig. 4.6   Notation for McLain's weighting function.

$\mathbf{x}_n$ of the subset of that tile for which $\mathbf{x}_i$ is the second-nearest data point. (The nearest data point is $\mathbf{x}_n$ by definition.)   Then $\lambda_n(\mathbf{x}_i)=0$ unless the tiles of $\mathbf{x}_i$ and $\mathbf{x}_n$ have a common side.   The functions λ_n can be extended to arbitrary points $\mathbf{x}$ by applying the definitions to the tessellation obtained by adding $\mathbf{x}$ to the data points.   A simple interpolator is obtained by using the λ_i's as the weights in a moving average, obtaining

$$f_0(\mathbf{x}) = \sum_n \lambda_n(\mathbf{x})Z(\mathbf{x}_n)$$

Sibson's preferred solution is to fit planes $f_1,\dots,f_N$ by weighted least squares, with weight $\{\lambda_i(\mathbf{x}_n)/d(\mathbf{x}_i,\mathbf{x}_n)^2\}$ for the value at $\mathbf{x}_i$ when fitting f_n. (These linear functions through each data point replace McLain's quadratic functions.)   Combining these functions with weights proportional to $\lambda_n(\mathbf{x})/d(\mathbf{x},\mathbf{x}_n)$ gives $f^*(\mathbf{x})$.   The factor $\lambda_n(\mathbf{x})$ gives this interpolator a local character, corresponding to McLain's sum over the vertices of the Delaunay triangle in which $\mathbf{x}$ lies.   The final interpolator is a constant linear combination of f_0 and f^* chosen to ensure that restricted quadratic surfaces of the form

$$a+bx+cy+d\left(x^2+y^2\right)$$

are interpolated exactly.

Akima (1978) fitted quintic (5th order) surfaces within each triangle. Such a surface has 21 coefficients.   Eighteen conditions are imposed to fit the surface and first and second derivatives $(Z, \partial Z/\partial x, \partial Z/\partial y, \partial^2 Z/\partial x^2, \partial^2 Z/\partial y^2, \partial^2 Z/\partial x\partial y)$ at the three vertices. The remaining three conditions are that the derivative of the surface perpendicular to each side should be a cubic function of the distance from the side.   In general, this would be a quartic, so this gives one condition per side.   A cubic is determined by the first and second derivatives at the two ends of the side, so this condition ensures that the derivative perpendicular to a side is continuous across the side.   Similarly, the value of the surface along a side is a quintic function of the distance along the side and so is determined by the values and first two derivatives along the side at the two vertices.   The net effect is to achieve a continuously differentiable surface within the triangulation.   The surface can be extrapolated beyond the convex hull of the data points as shown in Figure 4.7.   Within $DABE$ we take a surface that is quintic along the direction of the side AB and quadratic perpendicular to AB, defined by the 12 conditions at A and B, whereas in EBF we take a quadratic surface defined by the values along BE and BF, or, equivalently, by the six conditions at B.

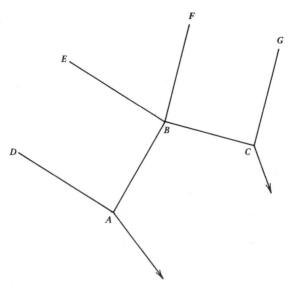

Fig. 4.7 Extending Akima's surface.

There are several ways to find the required first and second derivatives. It is possible that the first derivatives could be given; from drilling samples it may be possible to estimate the slope of a rock layer. Otherwise, one could fit a quadratic surface by distance-weighted least squares and use the derivatives of the fitted surface. Akima's preferred method is to take at each vertex nearby points in pairs and to form the normal to the plane through the vertex and the two points. These normals are averaged and the resultant vector is used to define the derivative plane at the vertex. Second derivatives are formed by applying the same technique to the estimated first derivatives.

Akima's method has no flexibility at all. It seems computationally reasonably fast and to yield pleasing surfaces. An algorithm is available on magnetic tape in the *CACM* series.

Powell and Sabin (1977) start from the premise that piecewise quadratic surfaces are good for contouring. Marlow and Powell (1976) had given a program that plots contours within a triangle by quadratic interpolation given values at the vertices and the midpoints of the sides. (This point is discussed in Section 4.5.) A triangle is divided into 12 subtriangles as in Figure 4.8. Then there is a unique surface, quadratic within each subtriangle with a continuous first derivative, with 12 parameters fixed by the surface values and slopes at the vertices (nine conditions) and the derivative perpendicular to the side at each of *P*, *Q*, and *R*. Such piecewise

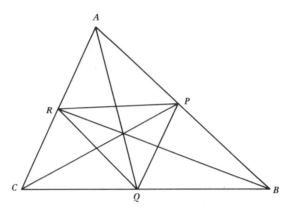

Fig. 4.8 Subtriangulation of Powell and Sabin.

quadratic surfaces are then continuous across a triangulation. (The reader is referred to the original paper for proofs.) The derivatives at the data points, if not given, can be estimated in any of the ways suggested for Akima's method. Powell and Sabin suggest taking the normal derivative at the midpoint as the average of the two normal derivatives at the two ends of the sides. This method again has no flexibility, and it is not clear how to extrapolate beyond the triangulation.

Akima (1978) refers to another piecewise quadratic method due to Whitten and Koellering (1975) that does not give a continuously differentiable surface.

It is not necessary to use a triangular grid, and so authors, including Hessing et al. (1972), have suggested forming an irregular rectangular grid from the data points. Such a grid can then be reformed into a regular rectangular grid by an affine transformation within each cell and the methods of Section 4.5 for interpolation from regular data applied. There is no known way to uniquely form this irregular grid. The techniques used by humans are too vague to program!

4.4 STOCHASTIC PROCESS PREDICTION

The analogue for spatial processes of Wiener–Kolmogorov prediction theory has been developed and used mainly by Matheron and his school in the mining industry. He christened the method "kriging" after D. G. Krige. This exposition takes a different point of view from the usual treatments; for example, those of Matheron (1965), Delfiner (1976), David

(1977), and Journel and Huijbregts (1978).   The approach and most of the evaluations are new, although foreshadowed by Whittle (1963b, Chapter 4).

The main distinction between this family of methods and those discussed so far is that the correlation between values of the surface at short distances is explicitly taken into consideration.   This remark also applies to values at data points, so the "weight" of each point in a cluster is automatically reduced.   Until further notice we assume that the covariance function $C(\mathbf{x}, \mathbf{y})$ is *known*; no stationarity condition is needed.   Conceptually we divide the surface $Z(\mathbf{x})$ into a trend and a fluctuation

$$Z(\mathbf{x}) = m(\mathbf{x}) + \varepsilon(\mathbf{x}) \qquad (4.14)$$

The distinction between the trend m and fluctuation ε is not clear cut. Generally we think of the trend as a large-scale variation, regarded as *fixed*, and the fluctuation as a small-scale *random* process.   Then C is the covariance function of ε as well as of Z.

A useful analogue is with time-series analysis, in which various methods are used to smooth the series, to remove the systematic component (often by methods analogous to those in the previous three sections).   This is equivalent to isolating our trend, which for some purposes may be the aim of the analysis.   In time-series research, a large amount of effort has been spent on describing the distribution of $\varepsilon(\mathbf{x})$ via correlations and spectra, on building parametric models for $\varepsilon(\mathbf{x})$ and on forecasting.

The importance of (4.14) is that we will predict the two components separately.   For a smooth summary we might use

$$\tilde{Z}(\mathbf{x}) = \hat{m}(\mathbf{x})$$

whereas for interpolation we use

$$\hat{Z}(\mathbf{x}) = \hat{m}(\mathbf{x}) + \hat{\varepsilon}(\mathbf{x})$$

The theory for the prediction of ε is precisely Wiener–Kolmogorov theory for a time series with a *finite* history.

Known Trend

Suppose first that $m(\mathbf{x})$ is known, so that we can work with $W(\mathbf{x}) = Z(\mathbf{x}) - m(\mathbf{x})$.   ($W$ is the ε process at present.)   We will consider only linear predictors

$$\hat{W}(\mathbf{x}) = \Sigma \lambda_i W(\mathbf{x}_i) = \lambda^T \mathbf{W}_N$$

and choose that which minimizes

$$E(W(\mathbf{x}) - \hat{W}(\mathbf{x}))^2 = \mathrm{var}(W(\mathbf{x})) - 2\Sigma\lambda_i C(\mathbf{x}_i, \mathbf{x}) + \Sigma\lambda_i\lambda_j C(\mathbf{x}_i, \mathbf{x}_j)$$

$$= \sigma^2(\mathbf{x}) - 2\lambda^T k(\mathbf{x}) + \lambda^T K \lambda \qquad (4.15)$$

where $K_{ij} = C(\mathbf{x}_i, \mathbf{x}_j)$, $k(\mathbf{x}) = (C(\mathbf{x}, \mathbf{x}_i))$, a column vector, and, of course, λ depends on $\mathbf{x}$. Now (4.15) is a quadratic form in λ that can be minimized by finding its stationary point. Differentiating with respect to each λ_i, in turn, we find

$$K\lambda = k(\mathbf{x}) \qquad (4.16)$$

$$\sigma_E^2(\mathbf{x}) = \min E(W(\mathbf{x}) - \hat{W}(\mathbf{x}))^2 = \sigma^2(\mathbf{x}) - \lambda^T k(\mathbf{x})$$

To solve (4.16) we need K to be nonsingular. Now K is a covariance matrix, and hence is nonnegative definite. It is invertible if and only if it is strictly positive definite, which we assume, and is no restriction in practice. Then our predictor becomes

$$\hat{W}(\mathbf{x}) = \left[\mathbf{W}_N^T K^{-1} \right] k(\mathbf{x}) \qquad (4.17)$$

$$\sigma_E^2(\mathbf{x}) = \sigma^2(\mathbf{x}) - k(\mathbf{x})^T K^{-1} k(\mathbf{x}) \qquad (4.18)$$

We interpret (4.17) as follows. The optimal (that is, minimum mean square error linear) predictor is that linear combination of the functions $C(\mathbf{x}, \mathbf{x}_i)$ that passes through the data points. Clearly, the optimal predictor at a data point $\mathbf{x}_i$ must be $Z(\mathbf{x}_i)$; (4.16) is solved by $\lambda_i = 1$, $\lambda_j = 0$ for all $j \neq i$. Thus $\hat{W}(\mathbf{x})$ interpolates W.

Computationally, (4.17) is convenient, for $\mathbf{W}_N^T K^{-1}$ is a row vector that can be formed once, after which $W(\mathbf{x})$ can be evaluated for each $\mathbf{x}$ with N additions and N multiplications. To form K^{-1} we use the Cholesky decomposition, which constructs a unique lower triangular matrix L with $K = LL^T$. Then if $y = K^{-1}\mathbf{W}_N$ we can find y by $L(L^T y) = \mathbf{W}_N$, easily in order N^2 operations since both L and L^T are triangular. To evaluate (4.18) we use

$$Le = k(\mathbf{x})$$

$$\sigma_E^2 = \sigma^2(\mathbf{x}) - \left(\sum_1^N e_i^2 \right) \qquad (4.19)$$

Note that we need order N^2 operations to form e and hence σ_E^2 for each $\mathbf{x}$. The most expensive operation is the Cholesky decomposition, which is needed only once and takes order N^3 operations.

Finding a Parametric Trend

Suppose we wish to find m, and we assume that it is a trend surface of the form $\mathbf{f}(\mathbf{x})^T\boldsymbol{\beta}$ discussed in Section 4.1. We now know the structure of the "errors" $\varepsilon(\mathbf{x}_1), \ldots, \varepsilon(\mathbf{x}_N)$ so we can use a more appropriate technique than least squares. Standard statistical theory shows that the covariance matrix for $\hat{\boldsymbol{\beta}}$ is minimized for the *generalized least-squares* estimate, which solves

$$\min (\mathbf{Z}_N - F\boldsymbol{\beta})^T K^{-1}(\mathbf{Z}_N - F\boldsymbol{\beta})$$

$$= \sum_{i,j} \left\{ Z(\mathbf{x}_i) - \mathbf{f}(\mathbf{x}_i)^T\boldsymbol{\beta} \right\} (K^{-1})_{ij} \left\{ Z(\mathbf{x}_j) - \mathbf{f}(\mathbf{x}_j)^T\boldsymbol{\beta} \right\} \quad (4.20)$$

using the notation defined at (4.3). Now (4.20) is

$$\min (\mathbf{Z}_N - F\boldsymbol{\beta})^T L^{-T}L^{-1}(\mathbf{Z}_N - F\boldsymbol{\beta}) = \min \| L^{-1}\mathbf{Z}_N - L^{-1}F\boldsymbol{\beta} \|^2$$

which reduces to a least-squares solution of the transformed problem

$$L^{-1}\mathbf{Z}_N = L^{-1}F\boldsymbol{\beta} + \boldsymbol{\eta} \quad (4.21)$$

Again, the lower triangular nature of L makes this an easy computation. The method automatically compensates for clusters of data points, for the values in a cluster will be highly correlated and so downweighted in the least-squares problem (4.21). The covariance matrix for $\boldsymbol{\beta}$ is

$$(F^T K^{-1}F)^{-1} = (F^T L^{-T}L^{-1}F)^{-1} = \tilde{R}^{-1}\tilde{R}^{-T}$$

for the reduction $\tilde{R}$ of $L^{-1}F$ found by the least-squares algorithm. Thus the variance of $\hat{m}(\mathbf{x}) = \mathbf{f}(\mathbf{x})^T\hat{\boldsymbol{\beta}}$ is given by $\| g \|^2$, where

$$\tilde{R}^T g = \mathbf{f}(\mathbf{x}) \quad (4.22)$$

Yet again we benefit from $\tilde{R}^T$ being a lower triangular matrix.

Prediction with an Unknown Trend

From our times-series analogy the obvious way to combine the last two subsections is to fit a trend surface $\hat{m}(\mathbf{x})$, form $W = Z - \hat{m}$, and then predict

W as if the trend had been known.　That is

$$\hat{Z}(\mathbf{x}) = \hat{m}(\mathbf{x}) + \hat{W}(\mathbf{x}) \tag{4.23}$$

$$\sigma_E^2(\mathbf{x}) = \mathrm{var}(\hat{W}(\mathbf{x})) + \mathrm{var}(\hat{m}(\mathbf{x})) \tag{4.24}$$

In fact, (4.23) gives the optimal predictor, but (4.24) is wrong. This is easily seen, for (4.24) predicts $\sigma_E^2(\mathbf{x}_i) > 0$ at a data point, whereas $\hat{Z}(\mathbf{x}_i) = Z(\mathbf{x}_i)$.

We need to modify our definition of an optimal predictor slightly. We look only at *unbiased* predictors for which

$$E(\hat{Z}(\mathbf{x})) = m(\mathbf{x})$$

From this

$$E(\hat{Z}(\mathbf{x})) = E(\Sigma \nu_i Z(\mathbf{x}_i)) = \Sigma \nu_i m(\mathbf{x}_i) = \Sigma \nu_i \mathbf{f}^T(\mathbf{x}_i)\boldsymbol{\beta}$$

should equal $\mathbf{f}^T(\mathbf{x})\boldsymbol{\beta}$ for all $\boldsymbol{\beta}$.　We deduce

$$\nu^T F = \mathbf{f}(\mathbf{x})^T \tag{4.25}$$

Then

$$E(Z(\mathbf{x}) - \hat{Z}(\mathbf{x}))^2 = \mathrm{var}(Z(\mathbf{x})) - 2\nu^T k(\mathbf{x}) + \nu^T K \nu \tag{4.26}$$

as before. To minimize (4.26) under condition (4.25) we introduce Lagrange multipliers μ and minimize

$$\mathrm{var}(Z(\mathbf{x})) - 2\nu^T k(\mathbf{x}) + \nu^T K \nu + 2(\mathbf{f}(\mathbf{x})^T - \nu^T F)\mu$$

This is again a quadratic form in ν, so we differentiate to find a stationary point and find

$$K\nu = k(\mathbf{x}) + F\mu$$

$$F^T \nu = \mathbf{f}(\mathbf{x})$$

$$\sigma_E^2(\mathbf{x}) = \mathrm{var}(Z(\mathbf{x})) - \nu^T k(\mathbf{x}) + \mu^T \mathbf{f}(\mathbf{x}) \tag{4.27}$$

We now show that $\nu^T \mathbf{Z}_N$ is $\hat{Z}(\mathbf{x})$, as defined at (4.23). The latter is clearly unbiased, for $E(\hat{m}(\mathbf{x})) = m(\mathbf{x})$, $E(\hat{W}(\mathbf{x})) = 0$, so we need only check the first condition of (4.27). Let $A = (F^T K^{-1} F)^{-1}$

$$\hat{m}(\mathbf{x}) = \mathbf{Z}_N^T K^{-1} F A \mathbf{f}(\mathbf{x})$$

$$\hat{W}(\mathbf{x}) = \mathbf{W}_N^T K^{-1} k(\mathbf{x}) = \mathbf{Z}_N^T \left[I - K^{-1} F A F^T \right] K^{-1} k(\mathbf{x})$$

$$\hat{Z}(\mathbf{x}) = \hat{m}(\mathbf{x}) + \hat{W}(\mathbf{x}) = \mathbf{Z}_N^T \left[K^{-1} k(\mathbf{x}) + K^{-1} F A \{ \mathbf{f}(\mathbf{x}) - F^T K^{-1} k(\mathbf{x}) \} \right]$$

We identify ν as $[\cdots]$ and can choose

$$\mu = A \{ \mathbf{f}(\mathbf{x}) - F^T K^{-1} k(\mathbf{x}) \}$$

to satisfy (4.27). The error variance becomes

$$\sigma_E^2(\mathbf{x}) = \left[\mathrm{var}(Z(\mathbf{x})) - k(\mathbf{x})^T K^{-1} k(\mathbf{x}) \right]$$

$$+ \left[\mathbf{f}(\mathbf{x}) - F^T K^{-1} k(\mathbf{x}) \right]^T A \left[\mathbf{f}(\mathbf{x}) - F^T K^{-1} k(\mathbf{x}) \right] \qquad (4.28)$$

The first term is $\mathrm{var}(\hat{W}(\mathbf{x}))$, but the second term is *not* $\mathrm{var}(\hat{m}(\mathbf{x})) = \mathbf{f}(\mathbf{x})^T A \mathbf{f}(\mathbf{x})$. The mean vector $\mathbf{f}(\mathbf{x})^T \boldsymbol{\beta}$ is adjusted to $[\mathbf{f}(\mathbf{x})^T - \lambda^T F] \boldsymbol{\beta}$, where λ is the vector of weights used to form $\hat{W}(\mathbf{x})$.

If $\mathbf{f}(\mathbf{x})$ is merely a constant (4.23) and (4.28) are the procedure Matheron calls "*kriging*" [although he considers the direct solution of (4.27)]. In this case, the unbiasedness condition is that the weights ν_i should sum to one (there is no reason why they should be nonnegative), which connects the method with the moving averages of Section 4.2. The general case is called "*universal kriging*." The following summarizes our computational procedure:

1. Form $K = [C(\mathbf{x}_i, \mathbf{x}_j)]$, L such that $LL^T = K$.

2. Form $F = \begin{bmatrix} f_1(\mathbf{x}_1) \dots f_P(\mathbf{x}_1) \\ \vdots \\ f_1(\mathbf{x}_N) \dots f_P(\mathbf{x}_N) \end{bmatrix}, \mathbf{Z}_N = \begin{bmatrix} Z(\mathbf{x}_1) \\ \vdots \\ Z(\mathbf{x}_N) \end{bmatrix}.$

3. Solve $L^{-1} \mathbf{Z}_N = L^{-1} F \boldsymbol{\beta}$ by least squares, reducing $L^{-1} F$ to $\tilde{R}$.

4. Form $\mathbf{W}_N = [Z(\mathbf{x}_i) - \mathbf{f}(\mathbf{x}_i)^T \boldsymbol{\beta}]$, y such that $L(L^T y) = \mathbf{W}_N$.

5. Predict $Z(\mathbf{x})$ by $y^T k(\mathbf{x}) + \mathbf{f}(\mathbf{x})^T \boldsymbol{\beta}$, $k(\mathbf{x}) = [C(\mathbf{x}, \mathbf{x}_i)]$

with error variance given by

$$\sigma_E^2 = \sigma^2(\mathbf{x}) - \|\mathbf{e}\|^2 + \|\mathbf{g}\|^2 \qquad \sigma_E^2 = \sigma^2(\mathbf{x}) - \mathbf{h}^T k(\mathbf{x}) + \|\mathbf{g}\|^2$$

$$L\mathbf{e} = k(\mathbf{x}) \qquad\qquad \text{or} \qquad L(L^T\mathbf{h}) = k(\mathbf{x})$$

$$\tilde{R}^T\mathbf{g} = \mathbf{f}(\mathbf{x}) - (L^{-1}F)^T\mathbf{e} \qquad\qquad \tilde{R}^T\mathbf{g} = \mathbf{f}(\mathbf{x}) - F^T\mathbf{h}$$

The first alternative is slightly faster, the second smaller if the least squares algorithm destroys $L^{-1}F$. In that case, $\mathbf{h}$ is the vector of weights.

Smoothness of the Surface

Suppose now that the covariance is homogeneous, so $C(\mathbf{x}, \mathbf{y}) = C(\mathbf{x} - \mathbf{y})$. The smoothness of the predicted surface is governed by the behavior of C for small $\mathbf{h}$, as shown by (4.17). We have seen that $Z(\mathbf{x})$ is always an interpolator, but this is achieved by a discontinuity at the data points if $C(\mathbf{h})$ is not continuous at zero. In that case we have what Matheron called a "nugget" effect, which could be attributed either to sampling errors in the data values or to very small-scale irregularities in the surface. The distinction between the two cases is as follows. If we repeat the measurement at exactly the same place we find the same value only in the second case, but if we sample close to the original point we obtain a substantially different value in both cases. With measurement errors we would want to ignore the discontinuous points on the predicted surface, which we can do by defining

$$k(\mathbf{x}_i) = \lim_{\mathbf{h} \to 0} C(\mathbf{h})$$

for each i. (This limit exists at least for isotropic covariances.) The realizations of a process with a nugget effect are very rough indeed, so we should not be surprised by discontinuities in the prediction. Let us assume that $C(\mathbf{h})$ is continuous at zero. The smoothness of the surface is then governed by the smoothness of $C(\mathbf{h})$ at zero. If

$$|C(0) - C(\mathbf{h})| \leqslant c|\mathbf{h}|^\alpha \qquad \alpha > 1 \qquad\qquad (4.29)$$

then $Z(\mathbf{x})$ is continuous. For a *Gaussian* process

$$|h|^{-\gamma} \sup_{|\mathbf{x} - \mathbf{y}| \leqslant h} |Z(\mathbf{x}) - Z(\mathbf{y})| \overset{a.s.}{\longrightarrow} 0 \qquad \gamma < \alpha/2 \qquad\qquad (4.30)$$

so the original surface is continuous if $\alpha > 0$, and n times differentiable if $\alpha > 2n$. [These results are given in one dimension by Neveu (1965), Proposition III.5.3 and Problem III.5.2. See also Cramér and Leadbetter (1968) Sections 4.2, 4.3, and 9.2. The multidimensional versions follow in exactly the same way.] From (4.20) $\hat{Z}(\mathbf{x})$ is continuous if $\hat{m}(\mathbf{x})$ and $C(\mathbf{h})$ are, and $\hat{Z}$ is n times differentiable if (4.29) holds with $\alpha > n$ (and $\mathbf{f}$ is sufficiently smooth). Thus the predictor $\hat{Z}$ will be rather smoother than the original surface even at the data points.

This interpolation approach is illustrated in Figures 4.9–4.11, for the data used in Figure 4.1. Figures 4.9 and 4.10 clearly illustrate the effect of the smoothness of $C(\mathbf{h})$ on the surface. Both have

$$C(\mathbf{x}, \mathbf{y}) = 1 - \exp - \{ d(\mathbf{x}, \mathbf{y}) / r_0 \} \tag{4.31}$$

with $r_0 = 5$ for 4.9, 20 for 4.10. The plotting and contour programs are not able to follow the sharp peaks! With the larger value of r_0 the predicted surface is able to follow the linear trend except at the edges. Figure 4.11 illustrates the prediction when a linear trend is specified and $r_0 = 5$.

Extensions

We can consider the prediction of the average of the surface over D or over blocks (for instance, the average ore grade in a section of a mine).

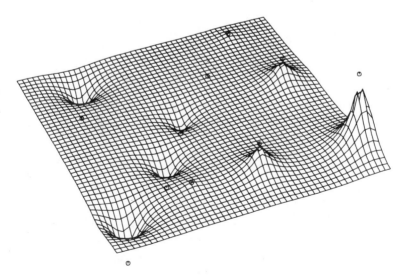

Fig. 4.9 Prediction of the surface from the same 10 data points as Figure 4.1, within a 100×100 unit square. Covariance given by (4.31) with $r_0 = 5$.

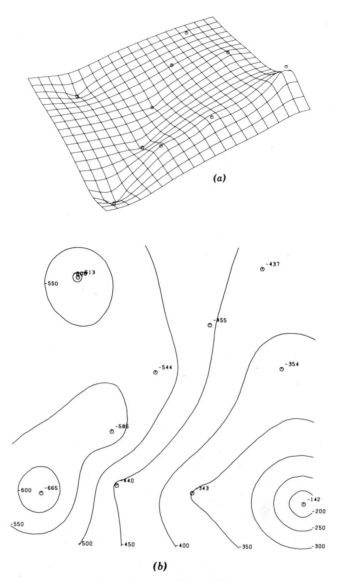

(a)

(b)

Fig. 4.10 As 4.9 with $r_0 = 20$.

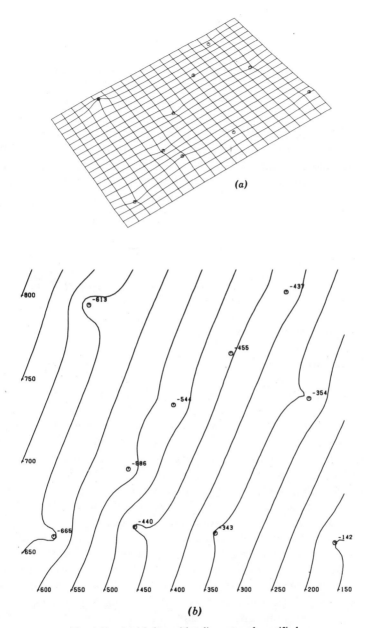

(a)

(b)

Fig. 4.11 As 4.9, but with a linear trend specified.

The only change will be to replace $k(\mathbf{x})$ by the vector of covariances between the variable to be predicted and the sample values, and $\mathbf{f}(\mathbf{x})$ by its average over the block. It is clear from (4.23) that the prediction will be precisely the average of the predicted surface over the block. This procedure is known as *block kriging* and is mathematically equivalent to following interpolation by a moving average. It can be used as a way of smoothing.

There is no reason why the data should be values at a point on the surface as we have assumed so far. They could for instance be averages over small blocks within D. Suppose that the ith observation is over the block B_i. Then we would need to replace K and $k(\mathbf{x})$ by the correct covariances:

$$k(\mathbf{x}) = \left[\int_{B_i} C(\mathbf{x}, \mathbf{y}) \, d\mathbf{y} / \text{area} \, (B_i) \right]$$

$$K_{ij} = \int_{B_i} \int_{B_j} C(\mathbf{x}, \mathbf{y}) \, d\mathbf{x} \, d\mathbf{y} / \{ \text{area} \, (B_i) \times \text{area} \, (B_j) \}$$

and $\mathbf{f}(\mathbf{x}_i)$ by $\int_{B_i} \mathbf{f}(\mathbf{y}) \, d\mathbf{y} / \text{area} \, (B_i)$.

Classes of Covariances

So far we have made the unrealistic assumption that the covariance function $C(\mathbf{x}, \mathbf{y})$ is known. It is this assumption that has saved us from assuming homogeneity. However, to estimate C we must make this assumption, and usually we will need to assume isotropy as well to obtain a sufficiently compact description of the covariance. If there are *a priori* grounds on which to assume that one direction differs in character from the others (such as depth in a three-dimensional description of a mine) we might assume *geometric anisotropy* and take $C(\mathbf{h})$ as a function of $\sqrt{(h_1^2 + h_2^2 + ah_3^2)}$, say, where a is an additional parameter.

Convenient classes of isotropic homogeneous covariances are rather few. Fortunately, what matters most in interpolation is the behavior of $C(r)$ near the origin, so we really only need to model the small-scale part of the covariance structure. However, we may well have *no* pairs of data points whose distance apart is in the critical range. This is a particular problem for a very regular pattern of sample points. In a mining or drilling situation we often will have dense sample points in a part of the study region. Even though this represents a "good" part of the region, we must assume that the local behavior there is typical. The prediction procedure essentially copies this local behavior elsewhere.

We have already used one parametric family of covariances, the exponentials at (4.31), and have shown in Section 2.2 that this is a valid isotropic covariance for all r in all dimensions. It produces a continuous but non-differentiable predicted surface.

Zubrzycki (1957) introduced a process found by counting the number of points of a Poisson process of intensity λ within distance $(R/2)$ of the sample point. Then

$$C(\mathbf{h}) = \lambda \operatorname{meas}(b(\mathbf{0}, R) \cap b(\mathbf{h}, R))$$

where $b(\mathbf{h}, R)$ denotes the ball of radius R centered at $\mathbf{h}$. This gives

$$C(r) = \begin{cases} \sigma^2 \left[1 - \dfrac{2}{\pi} \left[\dfrac{r}{R} \sqrt{\left(1 - \dfrac{r^2}{R^2}\right)} + \sin^{-1}\left(\dfrac{r}{R}\right) \right] \right] & r \leqslant R \\ 0 & r \geqslant R \end{cases} \qquad (4.32)$$

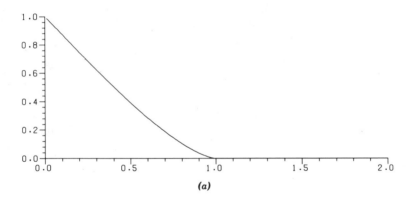

(a)

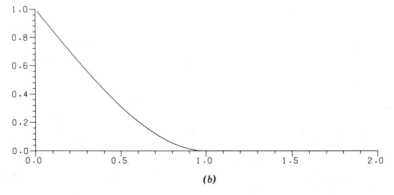

(b)

Fig. 4.12 Zubrzycki's correlation function in two dimensions (a) and three dimensions (b).

in two dimensions and

$$C(r) = \begin{cases} \sigma^2[1 - 3r/2R + r^3/2R^3] & r \leqslant R \\ 0 & r \geqslant R \end{cases} \qquad (4.33)$$

in three dimensions. Figure 4.12 illustrates these functions. Their behavior near the origin is very similar to an exponential function. Note that although the description is of a discontinuous surface, a Gaussian surface with this covariance is continuous by (4.30). The advantage of these models is that data points more than R from the point $\mathbf{x}$ being considered can be ignored, which can be used to reduce the computations.

Whittle (1954, 1963a) considered models with

$$C(r) = \frac{\sigma^2}{2^{\nu-1}\Gamma(\nu)} (r/r_0)^\nu K_\nu(r/r_0) \qquad \nu > 0 \qquad (4.34)$$

These functions are illustrated in Figure 4.13. The choice $\nu = \frac{1}{2}$ is the exponential class; for $0 < \nu < \frac{1}{2}$, $C(r)$ has an infinite derivative at 0, for $\nu > \frac{1}{2}$ a zero derivative there. The class

$$C(r) = \sigma^2 \exp\{-(r/r_0)^2\} \qquad (4.35)$$

also has a very smooth behavior for small r. (It is unfortunately sometimes called the "Gaussian" family!) We saw in Section 2.2 that these were valid covariances.

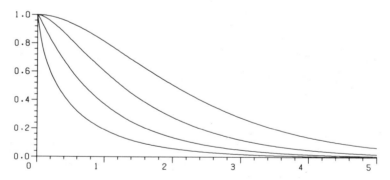

Fig. 4.13 Whittle's family of correlation functions. The values of ν are 2, 1, 0.5, 0.2 from top to bottom, all with $r_0 = 1$.

Fitting Covariances

In contrast to time-series analysis, very little work has been done on fitting the parameters of any of these models. We have seen that small distances are probably most important for interpolation. The usual approach is to look at the empirical covariances. To do so we form $c_{ij} = \{Z(\mathbf{x}_i) - \bar{Z}\}\{Z(\mathbf{x}_j) - \bar{Z}\}$ for all pairs of data points, and average all values for pairs $(\mathbf{x}_i, \mathbf{x}_j)$ with $\mathbf{x}_i - \mathbf{x}_j$ in each of a set of intervals. For the isotropic case—the only one we shall consider in detail—we average c_{ij} for pairs with $d(\mathbf{x}_i, \mathbf{x}_j)$ in each of the set of intervals. Figure 4.16a illustrates a plot of correlations in which the distances were divided into 100 equal-size intervals. Thus the point at r represents the average correlation between data points whose distance apart is in the interval $(r - \Delta/2, r + \Delta/2)$ for $\Delta \approx 0.09$ km. Of course some intervals of distance include no pairs of data points, and then no correlation is plotted. A problem with such plots is that the variability decreases as the distance increases, for the average number of pairs in each interval is proportional to the distance. The right-hand points are averages of many more points than those at small distances. Unfortunately, the latter interest us most. The correlation functions used in the examples have been fitted to such plots by eye. It is not always easy to decide when to include a "nugget effect" and allow a discontinuity in the correlation function at the origin, or whether to choose a linear (4.31 and 4.32) or quadratic (4.35, 4.34 for $\nu > \frac{1}{2}$) behavior at the origin.

In the absence of formal fitting procedures, the best way to assess the fit of a covariance function seems to be by *cross-validation*. Each data point is deleted in turn and its value predicted from the rest of the data, using the fitted covariance function. The prediction errors are then assessed. It is tempting to form the sum of squares of these errors. This may, however, not be suitable, for it may be dominated by a few data points that, because of their isolation, are hard to predict. Scaling the prediction errors by their standard errors and then forming a sum of squares biases the comparison in favor of covariances which give high standard errors by means of small covariance at typical interpoint distances. Even if a suitable summary could be found, its numerical minimization by altering the parameters of the models is probably too expensive for more than a few data points.

A complication arises when the covariance has to be assessed from data to which a trend surface is to be fitted, for the correlation plot should be of the residuals from the surface, which is to be fitted by generalized least squares using the correlation function. This apparent circularity can be broken by an iterative procedure. First fit a surface by least squares (or another guess at the covariance function), guess a covariance function, refit the surface using this, and refit a covariance function to the new residuals. If necessary, the process can be iterated. Forming correlations of residuals introduces a bias, for the correlation function of the residuals differs from that of the original ε process. The difference should be small at short distances if (4.14) is really a separation into a large-scale trend and small-scale fluctuation.

An Example

The data are 51 measurements of height of the earth's surface within a 310-foot square, from Davis (1973, Table 6.4). Figure 4.14 illustrates interpolation by weighted averages, with weighting functions d^{-1}, d^{-2}, and d^{-4}. The unsatisfactory nature of the first two choices is clearly seen. (Most of the small spikes up or down to the data points have been missed by the contouring program.) Figure 4.15 shows trend surface fits

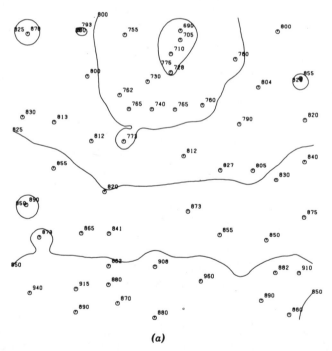

(a)

Fig. 4.14 Weighted averages for topographic data within a 310-foot square. Weighting functions were (a) d^{-1}, (b) d^{-2}, (c) d^{-4}.

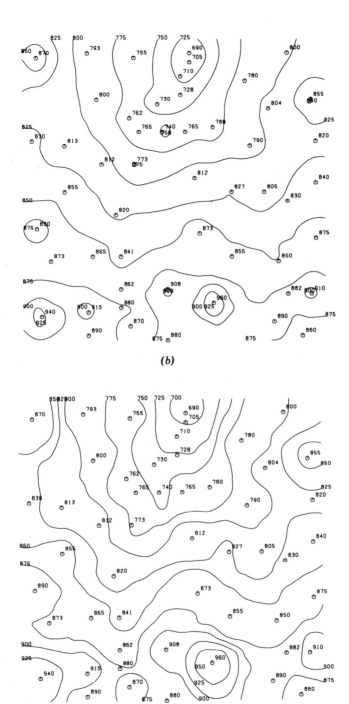

(b)

(c)

Fig. 4.14 (*continued*)

59

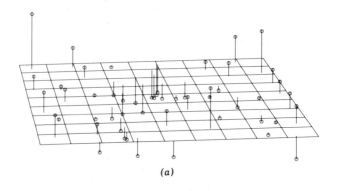

(a)

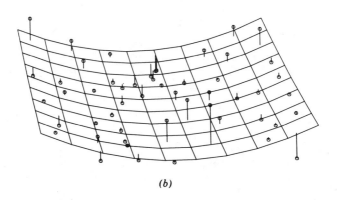

(b)

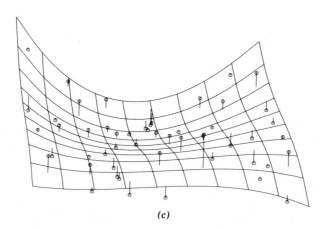

(c)

Fig. 4.15 Trend surfaces for topographic data (from Davis, 1973, Table 6.4) of orders 1(*a*) to 5(*e*).

60

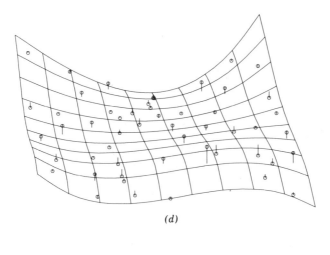

(d)

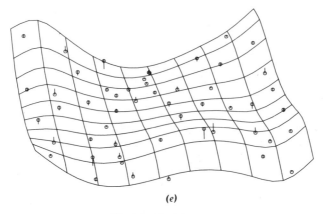

(e)

Fig. 4.15 (*continued*)

of orders 1 (linear) to 5 (quintic). The cluster of points at the middle back causes that part to be fitted rather too well. Figure 4.16 shows correlation plots for the data and residuals from second and fourth order surfaces, together with some covariance functions fitted by eye. The corresponding generalized least-squares surfaces are not shown, but differ only in that the cluster of data points is given less weight. Figures 4.17–4.21 show the predicted surfaces and the prediction standard errors for the five models fitted in Figure 4.16. Introducing a trend surface makes little difference to the prediction except at the corners, but does reduce the standard errors somewhat. The big difference is made, however, by the choice between (4.31) and (4.35), which have very different characters at the origin. The

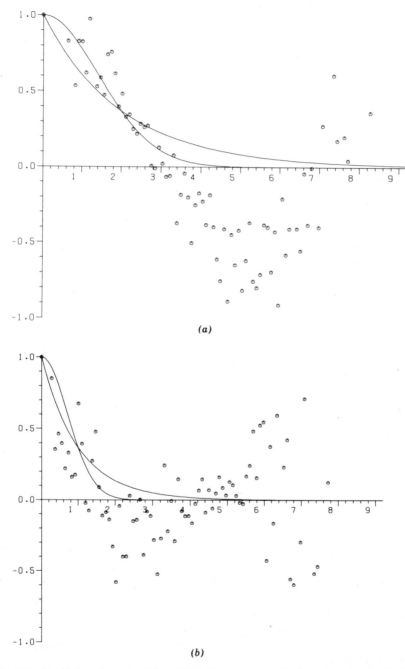

Fig. 4.16 Correlation plots. (*a*) Data. (*b*) Residuals from quadratic surface. (*c*) Residuals from quintic surface. The covariances shown are (*a*) (4.31) and (4.35) with $r_0 = 2$, (*b*) (4.31) and (4.35) with $r_0 = 1$, and (*c*) (4.31) with $r_0 = 0.5$. The abscissa is distance in units of 50 feet.

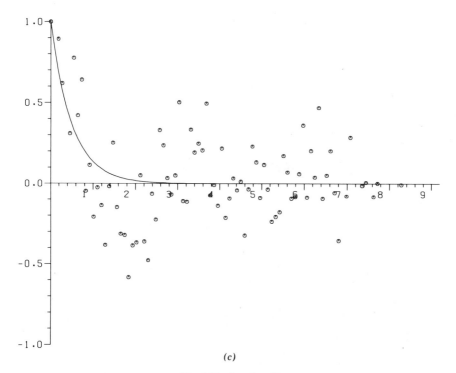

(c)

Fig. 4.16 (*continued*)

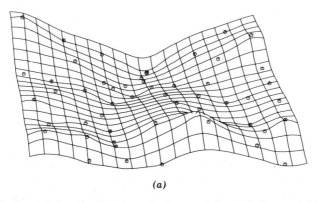

(a)

Fig. 4.17 Predicted surface (*a*, *b*) and standard error of the prediction error (*c*) for (4.31) with $r_0 = 2$.

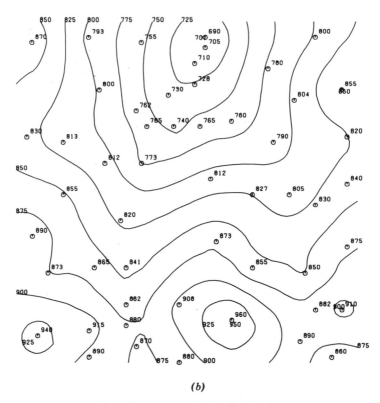

(b)

Fig. 4.17 (*continued*). Predicted surface.

quadratic behavior of (4.35) gives a much more "rounded" surface and much lower standard errors. The data give little help in making this choice, which needs external evidence.

Conditional Simulation

The prediction standard errors are little help in assessing the variability of nonlinear functions of the predicted surface, such as the area above a critical value, so it is helpful to have a means to simulate the fluctuations of the unknown parts of the surface. Suppose S is a simulation of a process with the covariance of the surface Z. Form a prediction $\hat{S}$ from

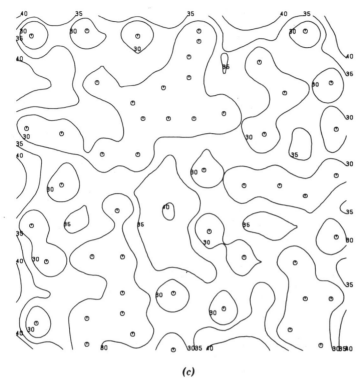

(c)

Fig. 4.17 (*continued*). Standard deviation of prediction error.

the values of S at the data points and

$$S(\mathbf{x}) = \hat{S}(\mathbf{x}) + \left\{ S(\mathbf{x}) - \hat{S}(\mathbf{x}) \right\}$$

$$Z(\mathbf{x}) = \hat{Z}(\mathbf{x}) + \left\{ Z(\mathbf{x}) - \hat{Z}(\mathbf{x}) \right\}$$

$$c(\mathbf{x}) = \hat{Z}(\mathbf{x}) + \left\{ S(\mathbf{x}) - \hat{S}(\mathbf{x}) \right\}$$

Then c is a realization of a surface with the same covariance as S and Z; $\hat{Z}$ and $\hat{S}$ are both interpolators, hence c passes through the data points. To

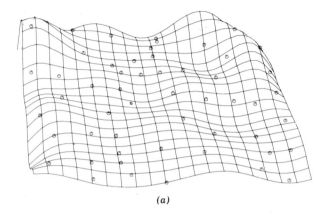

(a)

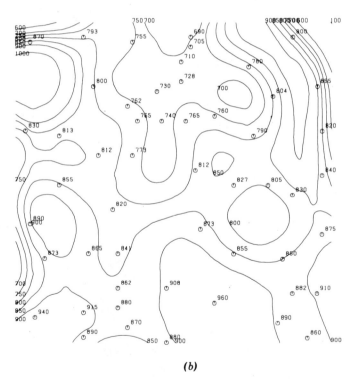

(b)

Fig. 4.18 As Figure 4.17 for (4.35) with $r_0 = 2$.

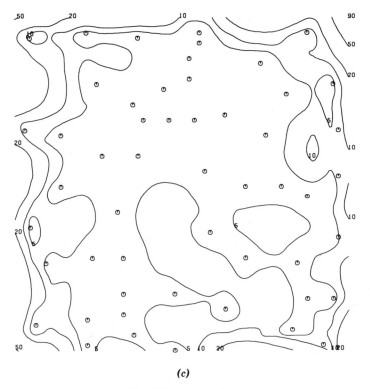

(c)

Fig. 4.18 (*continued*)

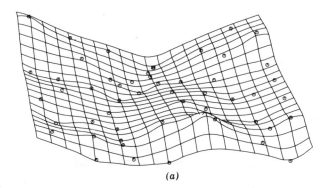

(a)

Fig. 4.19 As Figure 4.17 for $r_0 = 1$ with a quadratic trend.

67

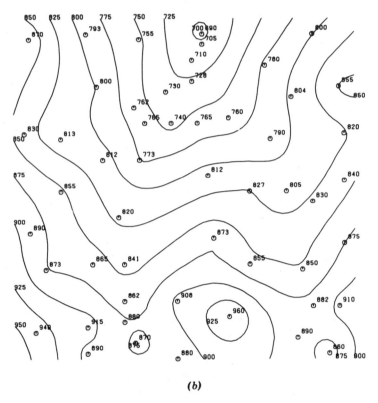

(b)

Fig. 4.19   (*continued*)

find the covariance function of c:

$$\text{cov}(c(\mathbf{x}), c(\mathbf{y})) = \text{cov}(\hat{Z}(\mathbf{x}) + S(\mathbf{x}) - \hat{S}(\mathbf{x}), \hat{Z}(\mathbf{y}) + S(\mathbf{y}) - \hat{S}(\mathbf{y}))$$

$$= \text{cov}(\hat{Z}(\mathbf{x}), \hat{Z}(\mathbf{y})) + \text{cov}([S - \hat{S}](\mathbf{x}), [S - \hat{S}](\mathbf{y}))$$

$$= \text{cov}(\hat{Z}(\mathbf{x}), \hat{Z}(\mathbf{y})) + \text{cov}([Z - \hat{Z}](\mathbf{x}), [Z - \hat{Z}](\mathbf{y}))$$

$$= \text{cov}(Z(\mathbf{x}), Z(\mathbf{y}))$$

The second equality follows from the independence of S and Z, the third from the equality of the distributions of S and Z, and the fourth from

$$\text{cov}(\hat{Z}(\mathbf{x}), [Z - \hat{Z}](\mathbf{y})) = 0 \qquad \text{for all } \mathbf{x}, \mathbf{y}$$

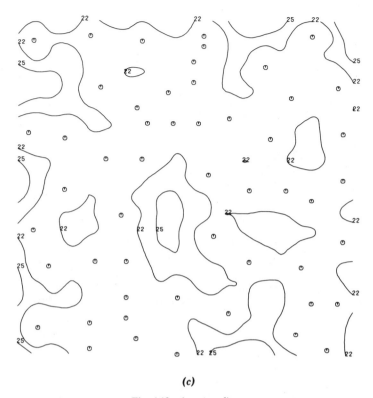

(c)

Fig. 4.19 (*continued*)

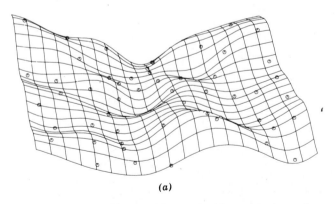

(a)

Fig. 4.20 As Figure 4.18 for $r_0 = 1$ with a quadratic trend.

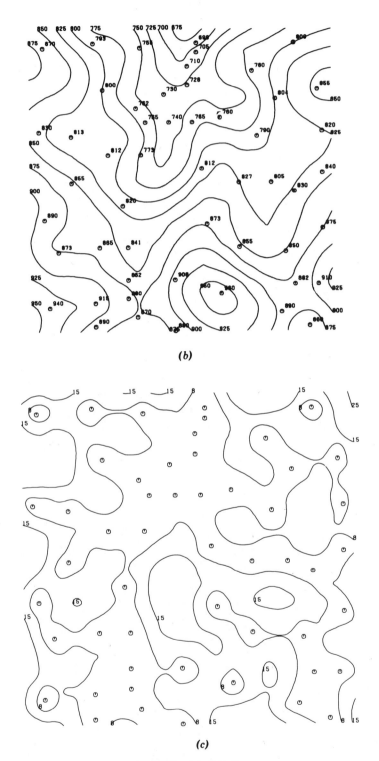

(b)

(c)

Fig. 4.20 *(continued)*

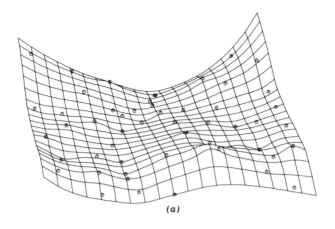

(a)

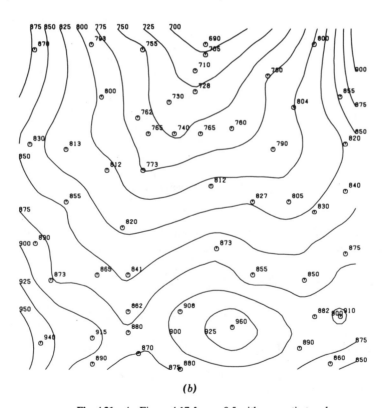

(b)

Fig. 4.21 As Figure 4.17 for $r = 0.5$ with a quartic trend.

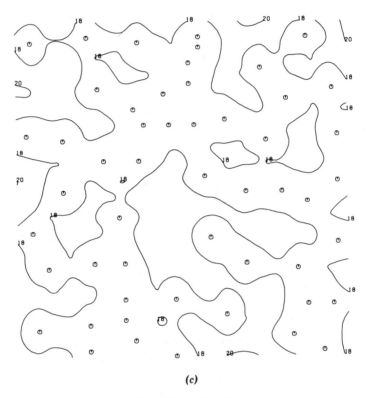

(c)

Fig. 4.21 (*continued*)

This is a characteristic property of an unbiased linear minimum mean-square error predictor and may be derived as follows:

$$E\big(Z(\mathbf{x}) - \hat{Z}(\mathbf{x}) + \alpha\hat{Z}(\mathbf{y})\big)^2 = E\big(Z(\mathbf{x}) - \hat{Z}(\mathbf{x})\big)^2$$

$$+ \alpha^2 E\big(\hat{Z}(\mathbf{y})\big)^2 + 2\alpha E\big(\{Z(\mathbf{x}) - \hat{Z}(\mathbf{x})\}\hat{Z}(\mathbf{y})\big) \quad (4.36)$$

The left-hand side of (4.36) is minimized by $\alpha = 0$, which can only be the case if

$$E\big(\{Z(\mathbf{x}) - \hat{Z}(\mathbf{x})\}\hat{Z}(\mathbf{y})\big) = \text{cov}\big(Z(\mathbf{x}) - \hat{Z}(\mathbf{x}), \hat{Z}(\mathbf{y})\big) = 0$$

because $\hat{Z}(\mathbf{x})$ is unbiased.

Nonlinear Problems

The minimum mean square error predictor of $Z(\mathbf{x})$ is always $E(Z(\mathbf{x})|\text{data})$. So far, we have only looked within the class of *linear* predictors. In general, this conditional expectation needs knowledge of the whole distribution of the process Z. However, for Gaussian processes conditional expectations are linear functions of the data, so minimum mean-square error predictors are linear and hence are found by the algorithms of this section. If we could find a transformation ϕ such that $\zeta(\mathbf{x}) = \phi^{-1}(Z(\mathbf{x}))$ is a Gaussian process, we would be advised to predict ζ and use $\phi(\hat{\zeta}(\mathbf{x}))$ as our predicted Z surface. Probably the only common example is when $Z(\mathbf{x})$ is thought to be lognormally distributed. Then defining

$$\zeta(\mathbf{x}) = \log Z(\mathbf{x})$$

each $\zeta(\mathbf{x})$ is Normally distributed, and it is a short step to assume joint Normality of $\zeta(\mathbf{x}_1), \ldots, \zeta(\mathbf{x}_N)$ and $\zeta(\mathbf{x})$, which is the definition of a Gaussian process. However, we need the mean to be relatively constant or to act multiplicatively, for if

$$\zeta(\mathbf{x}) = m(\mathbf{x}) + \eta(\mathbf{x})$$

then

$$Z(\mathbf{x}) = e^{m(\mathbf{x})} e^{\eta(\mathbf{x})} = e^{m(\mathbf{x})} \varepsilon(\mathbf{x})$$

where η is a Gaussian process and $\varepsilon(\mathbf{x})$ is lognormally distributed.

We could attempt to assess the distribution of $Z(\mathbf{x})$ by looking at a histogram of the data values. This may be confused both by the varying means (if a trend needs to be removed) and by the correlation between these random variables. Further research is needed.

Mining engineers are interested in estimates of the numbers of blocks in a mine with average grade above the economic cutoff point and in the total amount of ore in such blocks. If Z_V denotes the average of the surface within V, to give estimates of "probable" and "proven" reserves in an oil or gas field we need estimates of

$$P(Z_V \geqslant x | \text{data}) \tag{4.37}$$

In mining x is the critical grade, whereas in reserve evaluation we choose x to give conventional probability levels. We have estimates of the mean and variance of Z_V given the data, namely $\hat{Z}_V$ and $\sigma_E^2(V)$. In the case of a

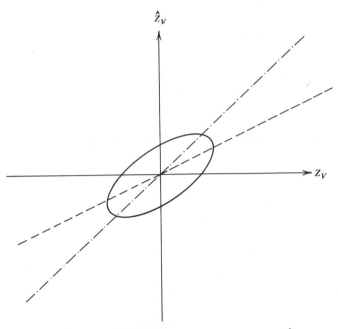

Fig. 4.22 Confidence ellipsoid for $\hat{Z}_V$ and Z_V. The regression lines for $\hat{Z}_V$ on Z_V(---) and Z_V on $\hat{Z}_V$(-·-) are shown.

Gaussian process (with known covariance structure), the conditional distribution of Z_V given the data is Normal with this mean and variance, so (4.37) may be estimated. However, whereas $E(\hat{Z}_V) = Z_V$, $\text{var}(\hat{Z}_V) <$ $\text{var}(Z_V)$ [cf. (4.18)]. A typical plot of $\hat{Z}_V$ vs. Z_V is given in Figure 4.22. An interesting point to note is that the regression of Z_V on $\hat{Z}_V$ is a line of unit slope, and for Gaussian processes this is $E(Z_V | \hat{Z}_V)$ (since $E(Z_V |$ data) depends on the data only through $\hat{Z}_V$). Thus whereas if the cutoff grade is above the mean, $\hat{Z}_V$ will be lower than Z_V, $\hat{Z}_V$ *does* provide a reasonable, unbiased predictor of Z_V, and the estimate of (4.37) should not be misleading.

These points are discussed in more detail by Matheron (1976a, b) and Maréchal (1976). Their "disjunctive kriging" can be viewed as a series expansion method of finding transformations to a Gaussian process and seems quite open to distortion by sampling variability.

References

Alldredge and Alldredge (1978) provide a collection of references, many of which are unpublished or privately published.

Agterburg and Chung (1973), Blais and Carlier (1967), Chiles and Matheron (1975), Dagbert and David (1976), David (1977, 1978), Delfiner (1976), Delfiner and Delhomme (1975), Dutta and Rao (1977), Haas et al. (1967), Huijbregts (1975), Journel and Huijbregts (1978), Krige (1966), Maréchal (1975, 1976), Maréchal and Serra (1970), Matheron (1963, 1965, 1967b, 1970, 1973, 1976a, b), Miesch (1975), Millier et al. (1972), Olea (1974), Venter (1976), Watson (1971a).

4.5 CONTOURING

Contouring algorithms take as input either an array of function values on a fine grid or a function that will evaluate the surface at any given point. If function evaluations are cheap, a technique that follows a contour, evaluating at points near to the contour as it goes, will be very accurate. The problem is to ensure that all the curves at a given contour level have been found. Since this method requires a large number of evaluations at each contour level, it is usually prohibitively expensive.

The alternative is to use a grid, usually rectangular. Contours are interpolated linearly along the edges of the grid and followed across the grid. To ensure all branches of the contour are found the boundary should be searched, then one set (horizontal or vertical) of internal edges. Previously traced branches can be found either by marking each edge visited or by checking an edge intersection against a list of all previously found points on contours at that level. Two problems arise when actually following contours. Figure 4.23 illustrates the three cases which can arise. At a saddle point [case (c)] a choice must be made. This can be done by taking the shorter of the total lengths of each pair of routes or, equivalently, by fitting the hyperbola $a + bx + cy + dxy$ within that square. The second problem arises when a contour actually passes through a grid point. This is avoided by perturbing downwards by a negligible amount all values very close to the contour level. An algorithm on these lines has been given by Snyder (1978). The contours in this book were drawn by a similar algorithm with the added refinement of passing a smooth curve

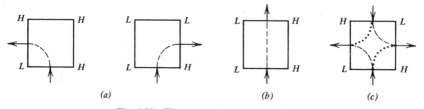

Fig. 4.23 Three cases in contour following.

through the edge intersections (cf. McConalogue, 1970, 1971). This refinement may cause contours to cross. For this reason, or because linear interpolation is not good enough, we may need to interpolate to a finer grid. The rest of this section discusses interpolation from data on a rectangular grid.

We can use the methods of the last three sections to interpolate within a rectangular grid, but simpler methods are available. We could use spline interpolation. The usual idea is to interpolate in one grid direction, and then the other, say from an $m \times n$ to $km \times n$ to $km \times ln$ grid. Then we can use one-dimensional spline interpolation. The choice of intermediate grid, $km \times n$ or $m \times ln$, does not matter (De Boor, 1962; Bhattacharyya, 1969, 1971; Whitten and Koellering, 1973).

There are simpler versions of the procedures of Akima and Powell. Akima (1974) uses a polynomial function in each cell of the grid, of the form

$$\sum_{r,s<3} a_{rs} x^r y^s$$

This is *not* a cubic polynomial, but is the form produced by a double application of cubic splines. However, Akima does not use the spline criterion of fit, but estimates $Z, \partial Z/\partial x, \partial Z/\partial y$ and $\partial^2 Z/\partial x \partial y$ at each corner of the cell and uses these 16 values to find the 16 coefficients. The resulting surface has a continuous first derivative. The proof is similar to that for his general scheme. The derivatives are estimated by local differences. Akima gives an example to support his claim that his method is preferable to cubic splines when contours run diagonally on the grid.

Powell uses the triangulation shown in Figure 4.24. Data are then available at six points on the boundary of each triangle, and a quadratic

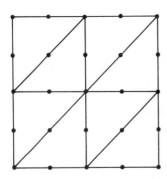

Fig. 4.24 Triangulation used in Powell's contouring algorithm.

surface is fitted within each, giving an overall surface with a continuous first derivative and piecewise elliptical contours. These are esthetically pleasing, but are produced in small pieces and very slowly (one-tenth the speed of the algorithm used for my figures). Usually some form of postprocessing will be necessary to follow a single contour from triangle to triangle.

CHAPTER 5

Regional and Lattice Data

Data on a regular lattice usually arise from a planned experiment or from a systematic sampling scheme. Agricultural field trials are carried out to compare varieties of crops. A field is divided into a number of rectangular plots (of the order of 400) separated by narrow strips, and varieties or treatments assigned to the plots by some experimental design. Clearly, we would expect a variation in fertility across the field that will be reflected as correlation in the "errors" of the linear model for yield associated with the experimental design. "Uniformity trials" were designed to investigate fertility variations. They are field trials with the same variety and treatment in every plot. One of the most famous is that reported by Mercer and Hall (1911). Fairfield Smith (1938) lists many others. Of course, the usual remedy is the use of blocks, as this was the experimental situation for which the analysis of variance was created. In Section 5.3 we investigate an alternative approach that models the correlations.

Other experiments performed on rectangular and triangular lattices are those designed to investigate the efforts of competition on plant growth. Typically young trees are planted at the lattice points, experimental designs being considered by Martin (1973) and Veevers and Boffey (1975). Lattice data also occur from systematically sampling a continuous surface (as advocated in Chapter 3) and from the quadrat sampling schemes of Sections 6.3 and 6.5.

Lattice data can be considered to be the two-dimensional analogue of a time series. This probably explains why it has received so much attention, for it is much rarer than data from irregularly spaced points or data on point patterns. In Section 5.1 we consider the multidimensional generalization of spectral analysis and in Section 5.2 parametric models related to autoregressive processes.

Regional data are sometimes regarded as occurring on an irregular lattice. Autoregressive models can be defined in a sufficiently general framework to be applied to observations on regions the connections

78

between which can be defined, for instance, by transport costs. This is done in Section 5.2, and the results applied in Section 5.4. Most of what geographers know as spatial statistical methods are multivariate rather than spatial in the sense of this volume, but the modeling of regression residuals by autoregressive processes provides a link with the rest of this chapter.

The reader is warned that Section 5.2 uses more matrix algebra and statistical theory than is needed elsewhere in this book.

5.1 TWO-DIMENSIONAL SPECTRAL ANALYSIS

Throughout this section we assume that we are given data (Z_{uv}) on a NX by NY rectangular grid with $N = NX \times NY$ observations in total. By analogy with time-series analysis, we form a *correlogram r* and *periodogram I* by

$$c(r,s) = \frac{1}{N} \sum_{m_1}^{M_1} \sum_{m_2}^{M_2} \left(Z_{uv} - \bar{Z} \right)\left(Z_{u+r, v+s} - \bar{Z} \right) \qquad (5.1)$$

$$m_1 = \max(1, 1-r) \qquad\qquad m_2 = \max(1, 1-s)$$

$$M_1 = \min(NX, NX-r) \qquad M_2 = \min(NY, NY-s)$$

$$r(u,v) = c(u,v)/c(0,0) \qquad (5.2)$$

$$I(\lambda, \mu) = \frac{N}{4\pi^2} \left| \frac{1}{N} \sum_{u,v} \left(Z_{uv} - \bar{Z} \right) e^{-i\lambda u} e^{-i\mu v} \right|^2 \qquad (5.3)$$

We can think of $c(r,s)$ as the average covariance over all pairs of observations whose coordinates differ by the vector (r,s). [There are $(NX - |r|)(NY - |s|)$ such pairs, suggesting this as a divisor in place of N. Alternative forms replacing $\bar{Z}$ by row and column means have also been used.] The functions c and r are symmetric in the sense

$$c(-r, -s) = c(r, s)$$

but

$$c(-r, s) \neq c(r, s)$$

in general. We will plot only the right half-plane; the left half-plane can be found by a half-turn rotation.

The periodogram has the same symmetry properties as the spectral density; we will plot the range $(-\pi < \lambda \leqslant \pi, -\pi < \mu \leqslant \pi)$. We can define I in real numbers by

$$I(\lambda, \mu) = \frac{1}{4\pi^2} \sum_{|u| < NX} \sum_{|v| < NY} c(u, v)\{\cos \lambda u \cos \mu v - \sin \lambda u \sin \mu v\}$$

Suppose that Z is a realization of a homogeneous stochastic process with mean m, covariance $C(\mathbf{h})$ and spectral density $f(\omega)$. Then if $\mathbf{h} = (r, s)$, $\mathbf{x} = (u, v)$, $\omega = (\lambda, \mu)$,

$$N E(c(\mathbf{h})) = \sum_{u, v} E\left[\{Z(\mathbf{x}) - m - (\bar{Z} - m)\}\{Z(\mathbf{x} + \mathbf{h}) - m - (\bar{Z} - m)\}\right]$$

$$= \sum_{u, v} \{C(\mathbf{h}) + \text{var}(\bar{Z})\} - E\left\{(\bar{Z} - m) \sum_{u, v} (Z(\mathbf{x} + \mathbf{h}) + Z(\mathbf{x}) - 2m)\right\}$$

$$\approx N\{C(\mathbf{h}) - \text{var}(\bar{Z})\} \tag{5.4}$$

when $|r| \ll NX$, $|s| \ll NY$. For large N we will usually also neglect $\text{var}(\bar{Z})$. Then (5.4) shows that $c(r, s)$ is an approximately unbiased estimate of $C((r, s))$ for small r and s. However, the sampling variance depends on $C(\mathbf{h})$ and neighboring values of the correlogram are substantially correlated. As in time-series analysis, the periodogram has a more satisfactory asymptotic sampling theory.

We say $\omega = (\lambda, \mu)$ is a *Fourier frequency* if λ and μ are multiples of $2\pi/NX$ and $2\pi/NY$, respectively. Then asymptotically (for large N) $I(\omega)/f(\omega)$ are independent at different Fourier frequencies with a standard exponential distribution, except at $(0,0)$, $(0, \pi)$, $(\pi, 0)$ and (π, π) ($I(0,0) = 0$ and $I(\omega)/f(\omega)$ has a χ_1^2 distribution at the other three frequencies). This result is a direct extension of the one-dimensional version (Bloomfield, 1976, p. 189). Thus for almost all Fourier frequencies ω

$$E(I(\omega)) \approx f(\omega) \qquad \text{var}(I(\omega)) \approx f(\omega)^2$$

So although $I(\omega)$ estimates $f(\omega)$, is is not a good estimator. Again borrowing ideas from time series we smooth $I(\omega)$ by a window W and use

$$g(\omega) = \int W(\Omega) I(\omega - \Omega) \, d\Omega \tag{5.5}$$

In our examples we have taken W as a bivariate Normal density

$$W(\omega) = \{\exp - \|\omega\|^2/2\alpha^2\}/2\pi\alpha^2 \qquad (5.6)$$

where α is an adjustable positive parameter. The approximate mean and variance of $g(\omega)$ are

$$E(g(\omega)) \approx f(\omega) \qquad (5.7)$$

$$\text{var}(g(\omega)) \approx 4\pi^2 f(\omega)^2 \int W(\omega)^2 \, d\omega/N = \pi f(\omega)^2/N\alpha^2 \qquad (5.8)$$

It would appear from (5.8) that we should take α as large as possible. However, the approximations are only good when $f(\omega)$ is reasonably constant over the range in which $W(\omega)$ is large (say over a radius of 2α in our example), which will limit the value of α. From (5.7) and (5.8) we see that taking logarithms will stabilize the variances, so we usually plot our estimate g of a spectral density on $\log_{10}$ scale. Then

$$\log_{10}g(\omega) = \log_{10}f(\omega) + \eta \qquad (5.9)$$

where η is usually taken to be approximately Normal with mean zero and variance

$$4\pi^2 \int W(\omega)^2 \, d\omega (\log_{10}e)^2/N = \pi(\log_{10}e)^2/N\alpha^2$$

The computation of a smoothed spectral density estimate is done using the Fast Fourier Transform (Bloomfield, 1976, Chapter 4). The mean is removed from the original array and this is extended by zeroes to an $NX2 \times NY2$ array, when each dimension is a power of 2. The *FFT* is applied to each row and then each column, the squared amplitude of each term in the resulting complex array taken, and the values scaled by $1/4\pi^2N$. This gives an array of values of I at frequencies of the form $(2\pi r/NX2, 2\pi s/NY2)$ for $0 \leqslant r < NX2$, $0 \leqslant s < NY2$. The estimate (5.5) is then approximated by a sum over these frequencies.

Mercer and Hall Data and Other Examples

Figure 5.1 illustrates the application of these methods to the wheat yields of the Mercer and Hall 1911 uniformity trial. There are 25 plots in the x direction (West–East) and 20 in the y direction (North–South), the plots

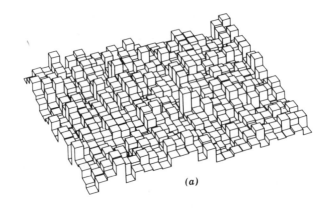

(a)

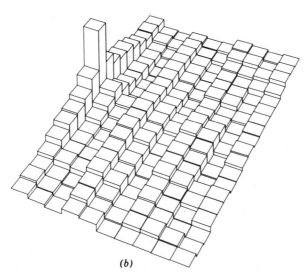

(b)

Fig. 5.1 Mercer and Hall's yields of wheat on a 20×25 grid of plots. (a) Perspective plot of the data. Left–right is West–East, front to back is South to North. (b) Correlogram. The spike is of unit height at the origin. Left–right corresponds to lags 0 to 8 in x, front to back to lags -8 to 8 in y. (c) Periodogram with range $(-\pi, \pi]$ in each direction. (d, e, f, g) Smoothed spectral density estimates on $\log_{10}$ scale. (d, e) are for $\alpha = 0.15$ and have 95% confidence range 0.75, (f, g) of $\alpha = 0.40$ with 95% confidence range 0.27.

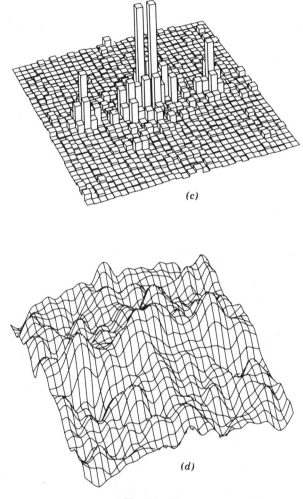

(c)

(d)

Fig. 5.1 (*continued*)

being about 11 feet square. The particular view of the data chosen illustrates its most marked feature, ripples running East–West. These show up as barely noticeable ripples on the correlogram and clearly on the periodogram as a peak at frequencies $(10\pi/32,0)$ and $(-10\pi/32,0)$, corresponding to a wavelength of about 35 feet in the East–West direction. Two different smoothings of the periodogram are shown with $\alpha = 0.15$ and 0.40. The width of the 95% confidence band at each frequency shows that

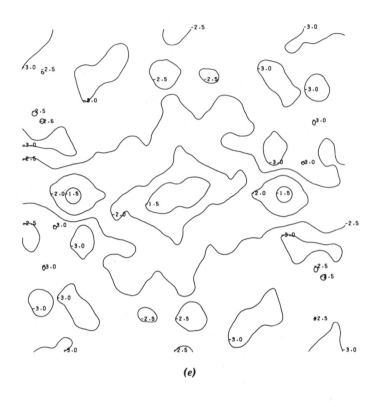

(e)

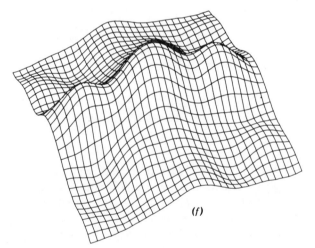

(f)

Fig. 5.1 *(continued)*

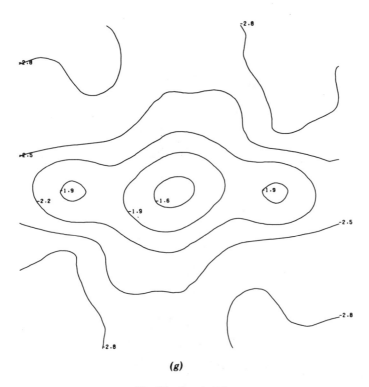

(g)

Fig. 5.1 (*continued*)

the two peaks are the only significant features of the less smooth picture, and we should increase the smoothing as shown in Figure 5.1 *f*, *g*.

Mercer and Hall noted the ripples in their data but attributed them to "casual irregularity." The spectral analysis shows that they *are* a significant feature of the data. Perhaps the most likely explanation is a variation in soil fertility caused by layers in the outcropping rocks. Patankar (1954) found an East–West linear trend representing an 11% change in yield across the field.

Satisfactory published examples of two dimensional spectral analysis seem rare. Rayner (1971) and Rayner and Golledge (1972) give some examples in geography, and Ford (1976) has a fascinating example from forestry research. A 120×36 meter plot of 40-year-old Scots pine trees was surveyed and the maximum crown height in each meter square recorded, as well as whether there was a trunk in that square. The longer axis of the plot was West to East, known to be the direction in which planting had taken place. To restrict attention purely to the canopy,

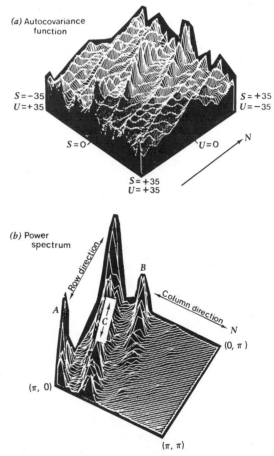

Fig. 5.2 Correlogram and estimated spectral density for crown heights of trees. (Reprinted with permission from Ford, 1976.)

heights were measured from 11.8 meters and negative heights truncated to zero. Figures 5.2 and 5.3 show the correlogram and a smoothed spectral density estimate from these truncated heights and the 1/0 matrix of presence/absence of tree trunks. The correlogram for crown height shows clear ridges at a spacing of about 18 meters in a direction about 15°W. Only one quadrant of the spectral density is shown (the figures in Ford, 1976 are incorrectly labeled). The full plot shows ridges with direction 75°E, parts of which are shown at A, B, and C in Figure 5.2b. The peak on this ridge at A corresponds to a wavelength of 2 meters in the

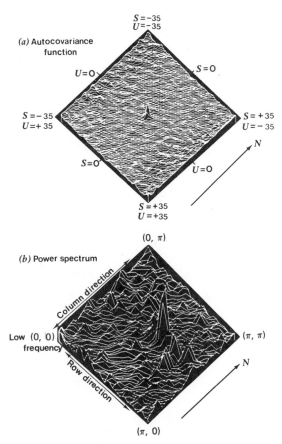

Fig. 5.3 Correlogram and estimated spectral density for the tree position matrix. (Reprinted with permission from Ford, 1976.)

East–West direction.   None of these features is evident in Figure 5.3; the peak at a diagonal frequency on Figure 5.3*b* is small enough to be attributed to chance.   No apparent pattern in the tree positions is not particularly surprising, for although the trees were planted in rows, about two-thirds will have been removed naturally or by thinning in a haphazard way.   The strong directionality found in the crowns could be attributable to prevailing wind directions or to a need for the crowns to expand diagonally given the original lattice planting.

Fasham (1978a, b) has used spectral analysis to study patterns of plankton.

5.2   SPATIAL AUTOREGRESSIONS

A simple autoregressive time series can be defined either by

$$X_t = aX_{t-1} + \varepsilon_t \qquad \varepsilon_t \sim N(0, \sigma^2) \text{ independent} \tag{5.10}$$

or by

$$E(X_t|\text{past values}) = aX_{t-1}, \qquad \text{var}(X_t|\text{past values}) = \sigma^2 \tag{5.11}$$

where X_t is assumed in the second case to be a Gaussian process. The asymmetry of conditioning on the past is important, for if we consider the two-sided versions

$$X_t = \frac{a}{2}(X_{t-1} + X_{t+1}) + \varepsilon_t \tag{5.12}$$

$$E(X_t|\text{rest}) = \frac{a}{2}(X_{t-1} + X_{t+1}), \qquad \text{var}(X_t|\text{rest}) = \sigma^2 \tag{5.13}$$

these define different processes and are prototypes for simultaneous (SAR) and conditional (CAR) autoregressions, respectively. The distinction between them is due to Brook (1964). Note that (5.12) is, in fact, the Yule process

$$X_t = b_1 X_{t-1} + b_2 X_{t-2} + \eta_t$$

in a nonstandard representation.

It will be most convenient to work in complete generality with observations $(Z_1, \ldots, Z_n)$ at n sites, following Besag (1975). We define SAR and CAR Gaussian processes by matrices S and C, assumed to have zeroes on their diagonals, and

SAR
$$Z_i = \mu_i + \sum S_{ij}(Z_j - \mu_j) + \varepsilon_i \tag{5.14}$$
$$\varepsilon_i \text{ independent } N(0, \sigma^2)$$

CAR
$$E(Z_i|Z_j, j \neq i) = \mu_i + \sum_j C_{ij}(Z_j - \mu_j) \tag{5.15}$$
$$\text{var}(Z_i|Z_j, j \neq i) = \sigma^2$$
$$\eta = Z - \mu - C(Z - \mu)$$

Let V_S and V_C be the corresponding covariance matrices for the (Z_i).

Theorem

$$E(Z_i) = \mu_i$$

$$V_S = \sigma^2 (I-S)^{-1} (I-S^T)^{-1} \tag{5.16}$$

$$V_C = \sigma^2 (I-C)^{-1} \tag{5.17}$$

$$\mathrm{cov}(\varepsilon) = \sigma^2 I \qquad \mathrm{cov}(\varepsilon, Z) = \sigma^2 (I-S^T)^{-1} \tag{5.18}$$

$$\mathrm{cov}(\eta) = \sigma^2 (I-C) \qquad \mathrm{cov}(\eta, Z) = \sigma^2 I \tag{5.19}$$

Necessary and sufficient conditions for existence are

SAR $(I-S)$ nonsingular.

CAR $(I-C)$ symmetric and strictly positive definite.

PROOF.   Let $Y = Z - \mu$

SAR $(I-S)Y = \varepsilon$ so $(I-S)E(YY^T)(I-S^T) = \sigma^2 I \tag{5.16}$

whence

$$\mathrm{cov}(\varepsilon, Z) = E(\varepsilon Y^T) = E\{\varepsilon \varepsilon^T (I-S^T)^{-1}\} = \sigma^2 (I-S^T)^{-1}$$

If $(I-S)$ is nonsingular, V_S is strictly positive definite. If $(I-S)$ is singular, $(I-S)Y = \varepsilon$ will, in general, have no solution.

CAR $(I-C)(V_C)_{ij} = E\left\{ \sum_k (\delta_{ik} - C_{ik}) Y_k Y_j \right\} = E\left(Y_i Y_j - Y_j \sum C_{ik} Y_k \right)$

$$= E\{ Y_i Y_j - Y_j E(Y_i | \mathrm{rest}) \}$$

$$= E\{ Y_i Y_j - E(Y_i Y_j | \mathrm{rest}) \} = E(Y_i Y_j - Y_i Y_j) = 0$$

for $j \neq i$

$$= E\{ Y_i^2 - Y_i E(Y_i | \mathrm{rest}) \} = \mathrm{var}(Y_i | \mathrm{rest}) = \sigma^2$$

for $i = j$

Thus $(I-C)V_C = \sigma^2 I$. If $(I-C)$ is singular, this has no solution. If $(I-C)$ is nonsingular $V_C = \sigma^2 (I-C)^{-1}$, so $(I-C)$ must be symmetric and strictly positive definite.

$$\text{cov}(\eta) = E(\eta\eta^T) = (I-C)E(YY^T)(I-C^T) = (I-C)V_C(I-C) = \sigma^2(I-C)$$

$$\text{cov}(\eta, Z) = E(\eta Y^T) = (I-C)E(YY^T) = \sigma^2 I \qquad\qquad \blacktriangledown$$

Notice that for a CAR scheme V_C determines C, whereas for an SAR process many matrices S can give V_S. The second formula in (5.19) is important, for it shows that CAR schemes have an "innovations" property generalizing the independence of X_t and $(\varepsilon_s, s > t)$ in (5.10).

Any SAR process is a CAR process with $C = S + S^T - S^T S$. Since we are considering only a finite set of sites, the reverse is true in a rather unnatural way. We can take $S = I - L^T$, where LL^T is the Cholesky decomposition of $(I-C)$.

The log likelihood of either scheme is given by

$$-\frac{n}{2}\ln 2\pi\sigma^2 + \frac{1}{2}\ln|B| - \frac{1}{2\sigma^2}(Z-\mu)^T B(Z-\mu) \qquad (5.20)$$

for

$$B = (I-S^T)(I-S) \quad \text{or} \quad B = I-C, \quad |B| \text{ the determinant of } B$$

If μ is specified by the linear model $D\theta$, then the maximum likelihood estimators of θ, σ^2 and parameters in B are given by

$$\hat{\theta} = (D^T\hat{B}D)^{-1}D^T\hat{B}Z \qquad (5.21)$$

$$\hat{\sigma}^2 = n^{-1}(Z-D\hat{\theta})^T\hat{B}(Z-D\hat{\theta}) \qquad (5.22)$$

which are the generalized least-squares solutions for covariance $\sigma^2\hat{B}^{-1}$, and $\hat{B}$ minimizes

$$-n^{-1}\ln|\hat{B}| + \ln\hat{\sigma}^2 \qquad (5.23)$$

For an SAR scheme $\hat{\sigma}^2$ is the mean square of the residuals

$$\hat{\varepsilon} = Z - D\hat{\theta} - \hat{S}\left[Z - D\hat{\theta}\right]$$

so that if the determinant $|B| = |I-S|^2$ is not identically one, (5.23) does not give the least-squares solution for parameters in S, and this solution may not be consistent (Whittle, 1954). (In Section 5.3 there is an example in which the least-squares solution is out by a factor of 2.) The problem in the numerical minimization of (5.23) is the evaluation of $|B|$ if this is necessary. Ord (1975) pointed out that if C is of the form ϕH for a single parameter ϕ, then

$$|B| = \prod_{1}^{n} (1 - \phi \xi_i)$$

where $\xi_1, \ldots, \xi_N$ are the eigenvalues of H. This formula also can be squared to give $|B|$ for an SAR with $S = \phi H$, but the eigenvalues may then be complex numbers if H is asymmetric.

Besag (1975, 1977c) introduced pseudo-likelihood estimation, in which the pseudo-likelihood

$$PL = \prod_{i} p(Z_i | Z_j, \, j \neq i)$$

is maximized. For a CAR process

$$\ln(PL) = -\frac{n}{2} \ln 2\pi\sigma^2 - \frac{1}{2\sigma^2} \|(I - C)(Z - \mu)\|^2 \tag{5.24}$$

so pseudo-likelihood estimation reduces to finding $\hat{\sigma}^2$ as the mean square of the residuals $\hat{\eta} = (Z - \hat{\mu}) - \hat{C}(Z - \hat{\mu})$ and choosing $\hat{C}$ to minimize $\hat{\sigma}^2$. The "innovations" property in (5.19) is the clue as to why a least squares fit is sensible for a CAR process, but not for an SAR process. In the simple case, $C = \phi H$, $\mu = 0$

$$\hat{\phi}^{PL} = Z^T H Z / \|HZ\|^2$$

$$\hat{\phi}^{PL} - \phi = \eta^T H Z / \|HZ\|^2 \tag{5.25}$$

and the innovations property gives the numerator of (5.25) a zero mean. As Besag (1975) points out, this allows ϕ^{PL} to be consistent if H varies with n in a suitable way.

The spatial nature of these autoregressions is introduced by taking the sites (now labeled by $\mathbf{r}, \mathbf{s}$) on a finite part of a rectangular lattice and parameterizing S or C using the lattice structure, so $S_{\mathbf{r}, \mathbf{s}}$ or $C_{\mathbf{r}, \mathbf{s}}$ depends only on the vector $\mathbf{s} - \mathbf{r}$. Some typical specifications are given in Figure 5.4. The scheme of Figure 5.4c is particularly simple as a CAR, for it is then equivalent to (f) as an SAR and to the product

$$Z_{uv} = X_u Y_v \qquad X_u = \lambda X_{u-1} + \varepsilon_u^{(x)}$$

$$Y_v = \nu Y_{v-1} + \varepsilon_v^{(y)}$$

for in each case $R(j,k) = \lambda^{|j|}\nu^{|k|}$, where $\alpha = \lambda/(1+\lambda^2)$, $\beta = \nu/(1+\nu^2)$. This process was introduced by Martin (1979).

Figure 5.4e, f, g illustrate one-sided schemes, which have the advantage that $|B| = |I - C|$ or $|I - S|^2$ is identically one and the computation of the maximum likelihood estimator is eased. The scheme of Figure 5.4g is adapted to a TV scan of a lattice, for it exhibits dependence only on past values of the scan. The other two are examples of the one-sided SAR schemes considered by Tjøstheim (1978) on an infinite lattice, for which $S_{r,s} = 0$ unless $r \geqslant s$ coordinatewise. Tjøstheim showed that direct analogues of the Yule–Walker estimation procedure and the Mann-Wald asymptotic theory for autoregressive processes are available for such schemes. They are easily re-expressed as symmetrical CAR schemes by $C = S + S^T - S^T S$. Moreover, Tjøstheim shows that any reasonable homogeneous process on an infinite lattice has an infinite one-sided simultaneous moving average representation. It is not easy to construct finite one-sided SAR approximations to other processes. Nevertheless, one-sided SAR schemes are important.

Whittle (1954) in a remarkable early paper gave some results as part of a general Fourier inversion method which amount to determining $|I - S|^{2/n}$ asymptotically for some simple schemes. For Figure 5.4b

$$-\frac{2}{n}\log|I - S| \to \sum_{j=1}^{\infty}\sum_{k=0}^{j}\frac{(2j)!}{[k!(j-k)!]^2}\alpha^{2k}\beta^{2(j-k)} \qquad (5.26)$$

and for $\alpha = \beta$ (Figure 5.4a) this reduces to

$$\sum_{j=1}^{\infty}\frac{1}{j}\binom{2j}{j}^2\beta^{2j} \qquad (5.27)$$

Other evaluations of $|B|$ based on finding eigenvalues are given by Besag (1977a, and in the discussion of Bartlett, 1978b), and Ord (1975, Appendix C).

"Coding" schemes for CAR processes use the distribution conditional on the observations at a carefully chosen set of sites (see Figure 5.5) so that

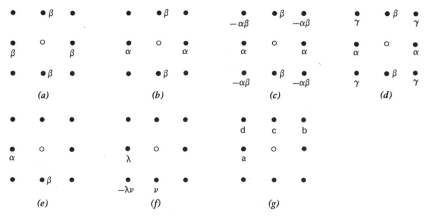

Fig. 5.4 Weights for autoregressive schemes. The value at the site marked by a circle is regressed on those sites with weights marked.

the remaining observations are conditionally independent with equal variances and means which are linear functions of the parameters, and are Normal. These parameters can then be fitted by least squares and have an exact conditional sampling theory. The procedure is wasteful of data and seems particularly likely to give estimates of $I - C$ that are not positive definite. Besag (1972b, 1974) used coding estimators; Besag and Moran (1975) and Besag (1977c) explored their efficiency and concluded that pseudo-likelihood estimators were preferable in the cases considered.

Even estimates of the asymptotic covariance matrix of the parameters in S or C are not readily available, and most of the published estimates are given without standard errors. Two exceptions are Tjøstheim's one-sided SAR schemes for which he gives the asymptotic theory [for Martin's process, the asymptotic covariance matrix of $(\hat{\lambda}, \hat{\nu})$ is n^{-1} diag $(1 - \lambda^2, 1 - \nu^2)$], and coding estimators (at least conditionally). If the likelihood is

```
•   ×   •   ×   •   ×   •

×   •   ×   •   ×   •   ×

•   ×   •   ×   •   ×   •

×   •   ×   •   ×   •   ×

•   ×   •   ×   •   ×   •
```

Fig. 5.5 One coding scheme for the process of Figure 5.4(b). Sites marked $\times$ are taken as the conditioning set.

maximized in part numerically some of the numerical second derivatives may be available. Although no general theory is available, there is a general belief that one will be developed paralleling that for Markov processes (Billingsley, 1961), which will show the usual relationship between these derivatives and the asymptotic covariance matrix. Whittle (1954) and Ord (1975) give some results for Gaussian processes.

Autoregressions have been fitted to the Mercer-Hall wheat yield data by Whittle (1954) and Besag (1974, 1977a). The periodicity noted in Section 5.1 may explain why the fit is not at all good. Furthermore, the data are not observations at lattice points, but represent averages of a continuous surface of soil fertility with some added measurement error on each plot. Besag (1977a) fitted a covariance matrix of the form

$$V = \alpha I + \beta (I - C)^{-1} \qquad \alpha, \beta > 0$$

corresponding to additive independent errors of variance α. Mead (1966, 1967, 1968, 1971) fitted SAR processes to small competition experiments.

Thus far we have confined attention to Gaussian processes. The conditional models can be viewed within the Gibbsian formulation of Section 2.4. Take the base process as independent $N(\mu, \sigma^2)$ observations. Then the probability density of a CAR process is

$$|I - C|^{1/2} \exp\left\{ \frac{1}{2\sigma^2} (Z - \mu)^T C (Z - \mu) \right\} \qquad (5.28)$$

If $C_{r,s} \neq 0$ only if r and s are neighbors on the lattice, then (5.28) is of the form (2.10) and the CAR scheme is both Markov and pair potential. Note that the normalizing constant is the awkward determinant.

This suggests a generalization by taking Markov pair potential processes with respect to other base processes. For binary variables Z_i we find

$$P(Z_i = z_i, i = 1, \ldots, N) = A \exp\left\{ \sum \alpha_i z_i + \sum \beta_{ij} z_i z_j \right\} \qquad (5.29)$$

$$P(Z_i = z_i | \text{rest}) = \frac{\exp\left\{ z_i \left(\alpha_i + \sum_j \beta_{ij} z_j \right) \right\}}{\left[1 + \exp\left\{ \left(\alpha_i + \sum_j \beta_{ij} z_j \right) \right\} \right]}$$

This is a logistic model for Z_i with the other observations as explanatory variables; Besag (1974) terms this an *autologistic* model. It is the classic

Ising model of statistical physics, in which the values 0 and 1 correspond to "up" and "down" magnetism of atoms in a crystal. Besag fitted this to presence/absence in a grid of quadrats (see Section 6.5), but as Mead pointed out in the discussion of that paper the model does not seem ecologically enlightening.

A further possibility is to take a count N_i at each site and modify a base process of independent Poisson variates. A simple pair potential process is given by

$$P(N_i = n_i, i = 1, \ldots, n) = A \exp\left\{ \sum_i [\alpha_i N_i - \ln N_i!] + \sum_{ij} \beta_{ij} N_i N_j \right\}$$

(5.30)

when the conditional distribution of N_i given the rest of the counts is Poisson with mean $\alpha_i + \sum_j \beta_{ij} N_j$. This Besag termed an *auto-Poisson* model. Considering $P(N_i = n_i, N_j = n_j | \text{rest})$ shows that A in (5.30) can be chosen to define a probability only if $\beta_{ij} \leq 0$ for all i, j. Thus an auto-Poisson process can only model competition within the population being counted.

Spatial moving averages, with V instead of V^{-1} parameterized in a simple way, are often mentioned but rarely used (Haining, 1978).

5.3 AGRICULTURAL FIELD TRIALS

One application of the spatial autoregressions discussed in Section 5.2 is to the field trials described in the introduction to this Chapter. We shall need only a few formulas from the previous section. Suppose that the p varieties have mean yields $(\theta_1, \ldots, \theta_p)$ and are distributed r times each according to some experimental design. Then if Z_i is the yield on the ith plot

$$E(Z_i) = (D\theta)_i \qquad D^T D = rI$$

Let N be the matrix of 0's and 1's giving which pairs of plots are neighbors. Then

$$Z = D\theta + \eta \qquad (5.31)$$

and we expect the "errors" η to be correlated. Papadakis (1937), in a method discussed further by Bartlett (1938, 1978a, b) proposed to adjust the variety means by an analysis of covariance involving the residuals from neighboring plots, that is, to fit

$$Z = D\theta + \alpha N \left[Z - D\hat{\theta}^{LS} \right] + \varepsilon \qquad (5.32)$$

where $\hat{\theta}^{LS}$ are the usual least-squares estimates. The analysis of covariance chooses α and θ by least squares. This is an approximation to the more logical idea of fitting

$$Z = D\theta + \alpha N [Z - D\theta] + \varepsilon$$

or

$$[I - \alpha N][Z - D\theta] = \varepsilon$$

or

$$Z = D\theta + \eta; \qquad \eta = \alpha N \eta + \varepsilon \qquad (5.33)$$

which introduces an SAR model for the residuals. From the theory of Section 5.2 the maximum likelihood estimator is given by

$$\hat{\theta} = (D^T \hat{B} D)^{-1} D^T B Z$$

$$D^T \hat{B} \left[Z - D\hat{\theta} \right] = 0 \qquad (5.34)$$

For our SAR process,

$$B = [I - \alpha N]^2 = [I - 2\alpha N + \alpha^2 N N],$$

so

$$r\hat{\theta} = D^T Z - \hat{\alpha} D^T N [2I - \hat{\alpha} N] \left[Z - D\hat{\theta} \right] \qquad (5.35)$$

The first term is the usual least-squares estimate. For small $\hat{\alpha}$ (5.35) is similar to (5.37) apart from the factor of 2. However, if we consider a CAR model for η in (5.31) with $C = \beta N$ we find

$$r\hat{\theta} = D^T Z - \hat{\beta} D^T N \left[Z - D\hat{\theta} \right] \qquad (5.36)$$

which is closer in spirit to Papadakis's estimator, for his estimate is given by

$$r\hat{\theta}^P = D^T Z - \hat{\alpha} D^T N \left[Z - D\hat{\theta}^{LS} \right] \qquad (5.37)$$

Indeed, if $\alpha = \beta$ were known, (5.37) can be regarded as a first step in the iterative solution of (5.36). Bartlett suggested the iteration of (5.37) and Martin in the discussion of Bartlett (1978b) showed that in one case the iterated solution of (5.37) converges to that of (5.36). It is perhaps not too surprising that a CAR appears when we note that least squares gives inconsistent estimates for an SAR process but is a reasonable (pseudo-likelihood) procedure for (5.31) with a CAR process of "errors" η. For small α, an SAR scheme with α and a CAR scheme with $\beta = \alpha/2$ give approximately the same covariance matrix. Since minimizing $n^{-1}\|(I - \alpha N)Z\|^2$ is consistent for the CAR scheme, least squares must give an estimate of approximately $\alpha/2$ for the SAR scheme! If we consider iterating (5.37) for fixed $\hat{\alpha}$ we find, with $E = \hat{\alpha}N/r$,

$$\hat{\theta}^{(i)} = r^{-1}D^TZ - D^TEZ + D^TED\hat{\theta}^{(i-1)}$$

$$= \left[I + D^TED + \cdots + (D^TED)^{(i-1)} \right] D^T \left[r^{-1}I - E \right] Z$$

$$+ (D^TED)^n \hat{\theta}^{(0)}$$

which converges to

$$\hat{\theta} = \left[I - D^TED \right]^{-1} D^T \left[r^{-1}I - E \right] Z$$

or

$$r\hat{\theta} = D^TZ - \hat{\alpha}D^TN \left[Z - D\hat{\theta} \right]$$

which is (5.36) with $\hat{\alpha} = \hat{\beta}$. For the convergence to hold, we need the largest eigenvalue of D^TND to be less than $r/\hat{\alpha}$. The matrix D^TND gives for each pair of varieties the number of times they were adjacent. Recently R. J. Martin showed that this condition holds for $\hat{\alpha}$ less than the reciprocal of the number of neighbors of each plot.

Adjusting by $\alpha N[Z - D\theta]$ is less attractive at the corner and edge plots that have fewer than four neighbors. It is not clear whether in practice it is better to adjust by the average of the (4, 3, or 2) residuals on neighboring plots or even to use diagonal neighbors at the corners. These suggestions correspond to alterations in the supposed covariance matrix of the "errors" η.

The reader is recommended to peruse Bartlett (1978b), especially the discussion, for details from both theoretical and practical points of view.

5.4   REGRESSION AND SPATIAL AUTOCORRELATION

The other application of spatial autoregressions has been in quantitative geography.   The observations are economic or population ·statistics on each of a set of contiguous regions.   The methods depend on a weighting matrix W expressing the connections between the regions.   This can be just a binary matrix giving 1 if the two regions have a common boundary, 0 otherwise, or it could depend on the length of the common boundary, the distances between the regions or transport costs between them (in which case W might not be symmetric).   Cliff and Ord (1973) and Haggett et al. (1977) discuss the choice of W at some length.   Bartels (1979) suggests that simple binary weights have proved as adequate as more complex schemes.   Even when W has to be estimated from other data it is regarded as fixed.

  The main thrust of research has been into investigating tests of no correlation among the observations or among residuals from a regression of the observations on explanatory variables.   Indeed, the philosophy adopted seems to have been that if "spatial autocorrelation" is found more explanatory variables should be introduced until it disappears!

  For binary weights Moran (1950) and Geary (1954) introduced the following coefficients of autocorrelation

$$I = \left[n \sum \delta_{ij}(x_i - \bar{x})(x_j - \bar{x}) \right] \bigg/ \left[\left(\sum \delta_{ij} \right) \sum_i (x_i - \bar{x})^2 \right]$$

$$C = \left[(n-1) \sum \delta_{ij}(x_i - x_j)^2 \right] \bigg/ \left[\left(2 \sum \delta_{ij} \right) \sum (x_i - x_j)^2 \right]$$

respectively, where $x_1, \ldots, x_n$ are the observations on n regions and $\delta_{ij} = 1$ if $i \neq j$ and i and j are contiguous, 0 otherwise.   Cliff and Ord generalized these definitions by replacing δ_{ij} by W_{ij}.   These coefficients have been thought of as tests of $\rho = 0$ in the SAR process

$$X = \mu + \rho W(X - \mu) + \varepsilon \tag{5.38}$$

For joint Normal data we can find the Neyman–Pearson test from (5.20) for $\rho = \rho_0 > 0$ as alternative to $\rho = 0$.   This is based on

$$\left[(x - \hat{\mu})^T (I - \rho_0 W)^2 (x - \hat{\mu}) \right] \bigg/ \left[(x - \bar{x})^T (x - \bar{x}) \right]$$

which for infinitesimal ρ_0 reduces to a multiple of I.

To discuss asymptotic Normality of I or C we have to consider how our set of regions might be supposed to become infinite. A proof of asymptotic Normality is given by Cliff and Ord (1973) and a more careful one by Sen (1976). The conclusion of these studies is that assuming Normality will not be a good approximation for sparsely connected regions such as those shown in Figure 5.6. A lot of effort has been spent in evaluating the theoretical moments of I and C both for independent Normal $x_1,\ldots,x_n$ and for the randomization of the given observations amongst the regions. We only consider independence. Cliff and Ord show that

$$E(I) = -1/(n-1)$$

$$E(I^2) = \left(n^2 S_1 - n S_2 + 3\Omega^2\right) / \left\{(n^2-1)\Omega^2\right\}$$

$$E(C) = 1$$

$$\mathrm{var}(C) = \left[(2S_1 + S_2)(n-1) - 4\Omega^2\right] / \left\{2(n+1)\Omega^2\right\}$$

where $\Omega = \Sigma w_{ij}$

$$S_1 = \tfrac{1}{2} \Sigma \left(w_{ij} + w_{ji}\right)^2$$

$$S_2 = \Sigma \left(w_{i\cdot} + w_{\cdot i}\right)^2$$

where $\cdot$ denotes summation over that index. These formulas are not valid for the residuals from a regression. Brandsma and Ketellapper (1979) note that there is no need to subtract the mean from residuals, and consider

$$Y = D\beta + \eta$$

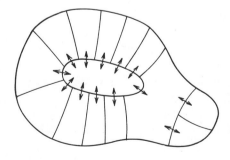

Fig. 5.6 A sparsely connected system for which spatial autocorrelation coefficients are not approximately Normal. Only regions joined by arrows are considered to be contiguous.

and test statistic

$$GMC = \frac{u^T W u}{u^T u} \qquad u = y - D\hat{\beta} \tag{5.39}$$

Then if β is fitted by least squares, $u = My$, where $M = [I - D(D^T D)^{-1} D^T]$, and

$$E(GMC) = tr\{MW\}/(n-k) \tag{5.40}$$

$$E(GMC^2) = \frac{tr\{MWMW^T\} + tr\{(MW)^2\} + \{tr(MW)\}^2}{(n-k)(n-k+2)} \tag{5.41}$$

where there are n observations and k explanatory variables and where $tr\{A\}$ denotes the trace of a matrix A (the sum of its diagonal elements). The key step in the derivation of all these formulas is the Pitman–Koopmans lemma, which asserts that the quotient and denominator of each of I, C, and GMC are independent for independent Normal observations.

Testing for spatial autocorrelation among the residuals of a regression is a slight extension of the problem of testing for serial correlation in the residuals of an econometric regression, and the recursive and *BLUS* procedures of that field have been tried. Brandsma and Ketellapper (1979) report a simulation study which recommended the *GMC* test statistics applied to the usual residuals, considered to be Normal with mean and variance given by (5.40) and (5.41). In particular, the full likelihood ratio test based on fitting β and ρ in

$$Y = D\beta + \eta \qquad \eta = \rho W \eta + \varepsilon$$

by maximum likelihood did not do well, it seems because of the inadequacy of the χ^2 approximation to the distribution of $-2\log(\text{likelihood ratio})$. (A systematic bias in this approximation appears in all their examples and will seriously affect the power comparisons.)

Cliff and Ord (1973) found a Normal approximation for both I and C to be adequate except for few regions ($n \leqslant 10$) or an unusual connections matrix W when the Monte Carlo tests of Section 2.5 should be used. They preferred I to C both in simulation studies and by showing that the asymptotic relative efficiency of C to I is not greater than one (as might be expected, since I approximates the Neyman–Pearson test), being $2S_1/(2S_1 + S_2 - 4n)$. However, the difference is small with regular systems of weights. Their Normality conclusions also apply to testing residuals by *GMC*.

The earliest studies of spatial autocorrelation were for presence/absence data. Suppose that at each site there is a color, black or white. The null hypothesis is that black occurs independently at each site with probability p. For binary weights Moran (1948) and Krishna Iyer (1949) evaluated the moments of the numbers BB of black–black and BW of black–white joins.

Lebart (1969) extended Geary's coefficient to give a spatial correlogram, using a series of weight matrices expressing "contiguities of different orders." Cliff et al. (1975) and Hodder and Orton (1976, pp. 179–183) give a correlogram in which the weights for each "lag" depend on the distances between regions.

The cited works of Cliff and colleagues and Dacey (1968), Fisher (1971) and Hordijk (1974) give applications of spatial autocorrelation in geography. Jumars et al. (1977), Jumars (1978), and Sokal and Oden (1978) give biological examples and Hodder and Orton (1976) give some uses in archeology.

A few attempts have been made to estimate the correlation structure and fit a linear regression model of the form

$$Y = D\beta + \eta, \qquad \eta = \rho W \eta + \varepsilon, \qquad \varepsilon \sim N(0, \sigma^2 I)$$

Bodson and Peeters (1975) fitted such a model by least squares, which as we saw in Section 5.2 is inconsistent. Ord (1975) and Hepple (1976) used maximum likelihood estimation of ρ and β simultaneously. The Cochran–Orcutt scheme used by Ord amounts to iterating (5.21), (5.22), and (5.23). Hepple (1979) gives a Bayesian approach (using the "natural conjugate prior" of ρ, $\log \sigma^2$ and β independent and uniform). The latter is very close in spirit to the approach of considering the whole likelihood function taken for example by Edwards (1972). Adrian Smith takes a different Bayesian view in the discussion of Bartlett (1978b).

It is possible to extend all this theory to space–time processes. If we assume say

$$Z(t) = \rho W Z(t-1) + \varepsilon(t)$$

where $Z(t)$ is a vector indexed by the regions, we do not encounter any of the problems posed by the determinant in (5.20). Such models are considered by Cliff and Ord (1975) and Cliff et al. (1975), and extensively by Bennett (1979). Even if W uses spatial structure these models are *multivariate* time series and have little in common with the methods of this volume.

CHAPTER 6

Quadrat Counts

In this chapter we look at analyses of data from small subareas, called *quadrats* after the square frames used by ecologists to mark out such samples. Sampling is usually either random or systematic and exhaustive as illustrated in Figure 6.1. One way to attempt random sampling has been to throw quadrats over one's shoulder; Greig-Smith (1964) reports studies that show that this is insufficiently chaotic, so the centers of the squares are usually found with reference to a table of random numbers. Within each quadrat the experimenter may count the population of plants or animals or take other measurements, such as the yield of a grassland or "cover." The latter is an estimate of the abundance of a plant species by further sampling; a number of pins are placed in the quadrat and the proportion that touch a plant of that species is recorded. Sampling in plant ecology is discussed by Greig-Smith (1964) and Kershaw (1973) and for animals by Southwood (1978).

6.1 INDICES

Quadrat sampling may have one or both of two distinct aims. A measure of the abundance per unit area is given by the average count for the quadrat samples divided by the quadrat area. Random positioning of the quadrats allows us to take averages over the randomization. Then the count per unit area from the samples is an unbiased estimator of the population count divided by the total area for the region within which the samples are taken. We will assume from now on that the measurement in each quadrat is a count and define the *intensity* to be the population count per unit area. Of course, the *variance* of this intensity estimate depends on the spatial pattern of the individuals being counted. If they are assumed to follow a Poisson process of intensity λ, the count within a quadrat of area A has a Poisson distribution with both mean and variance

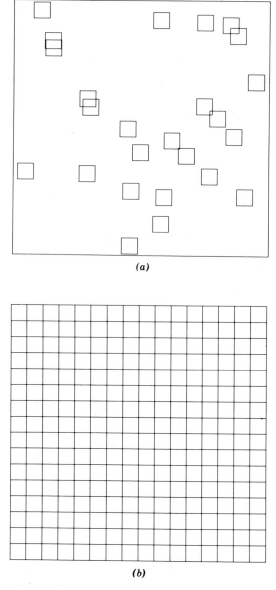

Fig. 6.1 (a) 25 random quadrats. (b) 16 × 16 grid of quadrats.

equal to λA. The unbiasedness of an intensity estimate from random quadrat sampling makes it a good method for estimating the total population size when it is feasible. Some practical problems and alternative methods are discussed in Chapter 7.

The other aim is to investigate the *pattern* of the population under study. Many indices have been proposed based on counts $(x_1, \ldots, x_n)$ from a set of quadrat samples. For Poisson-distributed counts, the variance/mean ratio is one. Fisher et al. (1922) and many others since have suggested considering the sample equivalent $s^2/\bar{x}$ (where $s^2 = \Sigma(x_i - \bar{x})^2/(n-1)$ and $\bar{x} = \Sigma x_i/n$ as usual). David and Moore (1954) introduced their "index of clumping"

$$ICS = s^2/\bar{x} - 1 \tag{6.1}$$

The idea of this index is that if the population is clustered, the index will be large, whereas if the individuals are regularly spaced the index will be negative. Furthermore, the sampling variability of the intensity estimate $\bar{x}$ will increase with ICS. The sampling distribution of ICS is unknown even for a Poisson process generating the population. For a significance test of the null hypothesis of a Poisson pattern $s^2(n-1)/\bar{x}$ is often referred to a $\chi^2_{(n-1)}$ distribution.

The notation ICS and $ICF = \bar{x}/ICS$ was introduced by Douglas (1975). ICS is his index of cluster size, for if we imagine a population with exactly m individuals at each point of a Poisson pattern, we would find $ICS \rightarrow (m-1)$ as the sample size increased and s^2 and $\bar{x}$ approached their means $m^2\lambda A$ and $m\lambda A$. If m is distributed as a Poisson random variable of mean μ independently at each point $ICS \rightarrow \mu$. The index of cluster frequency ICF should measure the mean number of clusters per quadrat and so be proportional to the area A of the quadrat, whereas ICS is (asymptotically in n) independent of the quadrat size or shape. For similar reasons, Lloyd (1967) defined an "index of mean crowding" x^* and "index of patchiness" IP by

$$x^* = \bar{x} + (s^2/\bar{x} - 1) = \bar{x} + ICS$$

$$IP = x^*/\bar{x} = 1/ICF + 1 \tag{6.2}$$

Lloyd considered x^* to represent the number of individuals sharing a quadrat with a typical individual, the two terms representing those in other clusters and those in the same cluster. If we look at the $n\bar{x}$ individuals in turn, we find the average number sharing the quadrat to be

$$\sum x_i(x_i - 1)/n\bar{x} = (1 - 1/n)(s^2/\bar{x} + \bar{x}) - 1$$

IP is found by rescaling by the intensity for a Poisson process.

Douglas (1975) derived standard errors for ICS and ICF as follows:

$$\left\{ICS - \left(\frac{\sigma^2}{\mu} - 1\right)\right\}^2 = \left[\frac{s^2}{\bar{x}} - \frac{\sigma^2}{\mu}\right]^2 = \left[\frac{\sigma^2 + (s^2 - \sigma^2)}{\mu + (\bar{x} - \mu)} - \frac{\sigma^2}{\mu}\right]^2$$

$$\approx \mu^{-2}\left[\sigma^2 + (s^2 - \sigma^2) - \sigma^2\frac{(\bar{x} - \mu)}{\mu} - \sigma^2\right]^2,$$

so

$$\mathrm{var}(ICS) \approx \mu^{-2}\mathrm{var}(s^2) - 2\sigma^2\mathrm{cov}(s^2, \bar{x})/\mu + \sigma^4\mathrm{var}(\bar{x})/\mu^2$$

The right-hand side can be expressed in terms of the first four moments of a count. These moments are then estimated by the moments of the observed counts. Douglas gives a computer program in APL to evaluate this and the comparable expression for ICF.

An alternative to the patterns of clusters at points of a Poisson process used so far is to assume patches, larger than a quadrat, of different intensities. Morisita (1959) introduced

$$I_\delta = n\sum x_i(x_i - 1)/\{n\bar{x}(n\bar{x} - 1)\} = n\bar{x} \cdot IP/(n\bar{x} - 1) \qquad (6.3)$$

The idea is that I_δ (or IP) will measure the variability in intensity between the patches and so be little affected by the quadrat size. If x_i has a Poisson distribution mean λ_i, the mean of I_δ converges for large $n\bar{x}$ to $n\sum\lambda_i^2/(\sum\lambda_i)^2$, so I_δ is a reasonable measure of the variability in λ.

Another justification used for these indices is to consider random thinning of the pattern, in which each individual survives (or is recognized) independently with probability θ. Then for large samples ICF is independent of θ, whereas ICS is reduced by the factor θ. (Proofs are given by Douglas, 1975 and Pielou, 1977, pp. 126–132.) It has been argued that random thinning only reduces the intensity but not the pattern features (Hill, 1973), but for our process of point clusters the clusters disappear for very small θ, since the probability that two or more individuals will survive from m is approximately $m(m-1)\theta^2/2$. Another counterexample is given by Ripley (1978, p. 971).

L. R. Taylor has for some years suggested another type of index for counts at distinct sites that could be applied to quadrats. He suggested that for a large number of data sets

$$s^2 \approx a\bar{x}^b \qquad (6.4)$$

and that b was a characteristic index. Taylor et al. (1978) give further references and check the fit of (6.4) to a large number of examples. Clearly, (6.4) must be an approximation, for at low intensities of counts

$$s^2 \sim \sum x_i^2/n - \bar{x}^2 \sim \bar{x} - \bar{x}^2$$

when the counts are 0 or 1. Nevertheless, the index b does seem to be a useful summary of the data without any specific spatial interpretation.

6.2 DISCRETE DISTRIBUTIONS

The methods of the previous section are not really spatial in that they ignore the spatial pattern of the sample. As such, they can really only be recommended for exploratory work and for their simplicity. Nevertheless, extensive research has gone into analyzing counts from random quadrats more closely and fitting parametric families of distributions to these counts. Rogers (1974) is a good reference at an elementary mathematical level. Douglas (1979) is more advanced.

Our two schemes of Section 6.1 can be formalized. The point cluster model gives rise to *generalized distributions* with

$$N = N_1 + \cdots + N_M$$

the N_i's being independent random variables all with distribution P_2, giving the number in each cluster, whereas M is an independent random variable with distribution P_1 giving the number of clusters. We refer to the distribution of N as $P_1 \vee P_2$. The patches of different intensity model gives rise to a *compound* or *mixed* distribution $P_1 \wedge P_2$ in which the parameters of a discrete distribution P_1 representing the count variability are sampled from a distribution P_2 representing the variations in intensity found by the random quadrat positioning.

Many distributions can be defined by one or both procedures. Common examples are

Negative binomial	= Poisson $\vee$ log series	= Poisson $\wedge$ gamma
Neyman type A	= Poisson $\vee$ Poisson	= Poisson $\wedge$ Poisson
Thomas	= Poisson $\vee$ (Poisson + 1)	

The index k of a negative binomial distribution has been suggested as an index in the sense of Section 6.1. The asymptotic value of ICF for a negative binomial distribution is k, and the methods of moments estimator of k found by equating $\bar{x} = E(N)$, $s^2 = \text{var}(N)$ is ICF. The maximum likelihood estimate of k can only be found by iteration or numerical maximization.

Consider the two explanations of the negative binomial distribution. For the patches model, we have as asymptotic values

$$ICF = \text{gamma shape parameter}$$

$$ICS = A/\rho \qquad\qquad A\text{-quadrat area}$$

$$\rho\text{-gamma scale parameter}$$

whereas for the point cluster model for a Poisson process of intensity ξ and log series parameter α

$$ICS \doteq \alpha/(1-\alpha)$$

$$ICF = A\xi/\{-\log(1-\alpha)\}.$$

Thus the variation in the two indices with quadrat size should discriminate between the two explanations. Unfortunately, both mechanisms could be present, and the mathematics is an unrealistic abstraction. To have any hope of understanding the spatial pattern several quadrat sizes will be needed.

Two of the underlying assumptions of this method are particularly suspect. Unless the quadrats are widely spaced the counts are unlikely to be independent as is assumed in the usual methods of estimating parameters. Indeed, in the patches model we must have no more than one sample per patch. Hence these methods should not be used for the exhaustive division into quadrats shown in Figure 6.1b. Second, it is unlikely that all the clusters will be small enough to be contained within a quadrat. Gleeson and Douglas (1975) explored in a simulation study the effect of realistically sized clusters when fitting the parameters of Neyman type A and Thomas distributions. (See also Pielou, 1957.) A related problem is to describe the association between two or more types of plants (different species or the same species in different years, for instance). Pielou (1977, Chapter 14) reduces the problem to a contingency table giving for types A and B the number of quadrats containing both A and B, only A, only B and neither. For r types an $r \times r$ table can be drawn up.

If this reduction in the data is acceptable and *if* the observations in different quadrats can be considered to be giving independent information then standard methods can be used to test for association in the table. Again, practically all the spatial information is ignored.

6.3   BLOCKS OF QUADRATS

Greig-Smith (1952) suggested the systematic division of an area into arrays of N quadrats, usually 16×16, as illustrated in Figure 6.1b. These basic quadrats can then be combined into larger rectangles as seen in Figure 6.2. Not only does blocking provide a sequence of quadrat samples of increasing area, it also uses the spatial arrangement of the data.   It must be noted that the area of the quadrats and the spacing of the information considered are increased at the same rate and effects due to large quadrats and to large distances could be confounded.   Kershaw (1957) proposed using a line of quadrats rather than a square grid.   His modification has been widely accepted because for a given amount of effort it provides information on many more "scales of pattern."   The difficulties of using a one-dimensional sampling technique to explore a two-dimensional pattern are perhaps more serious than users of Kershaw's method have suggested. The following discussion applies to either lines or grids.

Greig-Smith's analysis was to compute the sum of squares S_r for each block area r and use these in a nested analysis of variance.   Consider S_r and S_{2r}.   The blocks at size $2r$ are each divided into two counts (a_i, b_i) for $i = 1, \ldots, m = N/2r$.

$$S_r = \frac{1}{r} \sum_1^m \left(a_i^2 + b_i^2 \right) \qquad S_{2r} = \frac{1}{2r} \sum_1^m \left(a_i + b_i \right)^2$$

$$S_r - S_{2r} = \frac{1}{2r} \sum_1^m \left[2a_i^2 + 2b_i^2 - a_i^2 - b_i^2 - 2a_i b_i \right] = \frac{1}{2r} \sum_1^m \left(a_i - b_i \right)^2 \quad (6.5)$$

We call S_r a sum of squares, not a mean square, as it is the reduction in sum of squares obtained by allowing a different mean in each of the $2m$ blocks of size r.   In a nested analysis of variance the mean square $M_r = (1/m)(S_r - S_{2r})$ is associated with allowing different means in each half of each $2r$ block.   Equation (6.5) shows that M_r is the average over blocks of size $2r$ of the squared differences between the counts in the two halves of the block divided by the block size $2r$.   It would therefore be more appropriate to label this mean square by $2r$, but we will keep to the

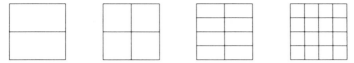

Fig. 6.2 Blocking of 16 quadrats into 4×2, 2×2, 2×1, and 1×1 blocks.

conventional terminology.   Hill (1973) proposed a "two-term local variance" in which all blocks of a given size and shape were used in the
average for M_r, not just the nonoverlapping set shown in Figure 6.2.

It is usual to plot M_r against r and to note peaks and troughs in the
resulting plot.   This is more satisfactory for a line than a grid, when
alternate points refer to square and rectangular blocks.   The usual F tests
in the analysis of variance could be used, but typically M_1 will be inflated
by clustering on the scale of a quadrat (which is biologically uninteresting).
Thompson (1955) calculated expected mean squares for Neyman-Scott
cluster processes (see Section 8.4); these he used in Thompson (1958) to
explore which features of the plots would correspond to parameters in
these models.   The assumptions of Normality and independence underlying the F distribution are usually untenable and "significance" has been
assessed by comparing plots from several transects and extracting their
common features.   It is important to note that the mean squares at larger
block sizes are based on fewer blocks and so are more variable.   Intrinsi-
cally there is less information at these block sizes and Hill's modification
cannot help.   Usher (1969, 1975) and his student Errington (1973) have
tested Greig-Smith's analysis for various artificial patterns and have shown
that the choice of the origin of a transect (and presumably also of a grid)
can markedly affect the plots.   A simple explanation is as follows.   Suppose there is a soil fertility fluctuation of period about $2r$ times the quadrat
size.   Then for some starting positions a large mean square will appear at
block size r and for some at $r/2$, as shown in Figure 6.3.

Moellering and Tobler (1972) used a similar analysis for a geographical
example.   Ecological applications are given by Allen (1973), Anderson
(1961, 1967, 1971), Austin (1968), Cooper (1960), Greig-Smith (1961, 1964),
Greig-Smith and Chadwick (1965), Hall (1971), Hill (1973), Kershaw
(1957, 1958, 1959, 1960, 1961, 1973), Morton (1974), Owen and Harberd
(1970), Pemadasa et al. (1974), Riechert (1974), Riechert et al. (1973),
Usher (1975), Walker et al. (1972), Westman (1975), Westman and Anderson (1970), and Williamson (1975).   Greig-Smith (1979) gives a recent
survey.

The data are counts so we could replace the analysis of variance by a
series of likelihood ratio tests for independent Poisson counts.   This

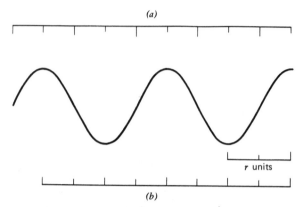

Fig. 6.3 The effect of the starting point of a transect, where (*a*) has a large mean square at block size *r* and (*b*) at size *r/2*.

analysis would correspond to a Poisson pattern allowed different intensities within the blocks. Nelder and Wedderburn (1972) term this an *analysis of deviance*. Let μ_j be the mean for the jth quadrat, and $m_j = \ln \mu_j$. Then the log likelihood of counts $Z_1, \ldots, Z_N$ is

$$L(m) = \sum_1^N \left\{ -\mu_j + Z_j m_j - \ln(Z_j!) \right\} \tag{6.6}$$

If we consider blocks of size r within blocks of $2r$ we have the test statistic T_r, which in the notation at (6.4) is

$$T_r = \sum_1^m (a_i + b_i) \ln\{(a_i + b_i)/2r\} - a_i \ln\{a_i/r\} - b_i \ln\{b_i/r\}$$

$$= L_{2r} - L_r - \left(\sum_1^N Z_i \right) \ln 2 \tag{6.7}$$

where L_r is the sum of $Z \ln Z$ for the total count in each block of size r. From standard maximum likelihood theory the distribution of $-2T_r$ is (independent) $\chi^2_{(N/2r)}$ asymptotically for large N. Orloci (1971) also derived (6.7), but from the standpoint of minimum discrimination information. He considered confidence bands based on randomizing the spatial positions of the N quadrat counts.

These Poisson likelihood ratio tests can be computed (rather inefficiently) by the program GLIM (Baker and Nelder, 1977), which can also be used to test other hypotheses. For instance, it could fit a trend surface model for m_j to reveal or remove broad trends in the counts. The assumption of a Poisson distribution for the counts is probably better than a Normal distribution, but the possible presence of clusters suggests that a generalized distribution would be better. GLIM can be used to assume only that the variance of the distribution is proportional to the mean, which is appropriate for such models. For the model with different means in blocks (but not for trend surface models) the effect is to multiply the deviance by an unknown constant. Ratios of the deviances again give F tests which will be more appropriate than those under Normality assumptions. Some examples are discussed in the next two sections.

Mead (1974) proposed tests based on randomizing the positions of the quadrats in a transect. Consider blocks in sets of four. For each four consider all possible differences between the totals of two pairs from the four. There are three such pairs. The observed difference between the sums of the first and second pair is compared with the randomization distribution. An example may be helpful. Counts

give
$$
\begin{array}{cccc}
0\ 2\ 2\ 0 & 0\ 0\ 1\ 10 & 11\ 1\ 0\ 2 & 5\ 9\ 4\ 10 \\
(0,0,4) & (11,9,9) & (10,8,12) & (0,10,2)
\end{array}
$$

for the differences between the halves, in each case the first of the three being the observed difference and the other two equally likely under randomization. The sum S is 21; under randomization $P(S \le 21) = 28/81$ and $P(S \ge 21)$ is $67/81$. A "distribution-free" form of the test is formed by replacing triples of differences of the form

$$
\begin{array}{llll}
(a,a,b) & \text{by} & (0,0,2) & a < b < c \qquad (6.8) \\
(a,b,c) & \text{by} & (0,1,2) & \\
(a,b,b) & \text{by} & (0,2,2) &
\end{array}
$$

If the counts are blocks of size r, this test is of the difference between the halves of a block of size $4r$ and so is labeled by $2r$. The virtue of this procedure is that the tests made at sizes $2,\ldots, N/4$ are *independent*. Although Mead did not do so, each of the t tests should be performed at level $1 - (1 - \alpha)^{1/t} \approx \alpha/t$ to obtain an overall significance level of α. The exact permutation distribution can be found for transects of up to 1024 quadrats by careful programming (to avoid overflows). For small block sizes it may be convenient to use a Normal approximation to the distribution of S. For the "distribution-free" version this is

$$
Z = \frac{S - (2n_1 + 3n_2 + 4n_3)/3}{\sqrt{\{8(n_1 + n_3) + 6n_2\}/3}}
$$

where there are n_1, n_2, and n_3 triples of the three types in (6.8). A two-dimensional analogue is available and is considered briefly by Mead (1974, p. 306) and Besag and Diggle (1977). An exact analogue can be applied to alternate horizontal and vertical pairings or only square blocks considered, so that four blocks of size $4r$ are considered within a $16r$ block. Any statistic that compared the four subblocks would have a permutation distribution over $(16!)/(4!)^5 \approx 2.6$ million permutations. Besag and Diggle suggested using random permutations to explore this distribution. One possibility is to use these random permutations to estimate the mean and variance of the statistics for each $16r$ block and refer the sum over the blocks to a Normal distribution.

Zahl (1974, 1977) looked for a method to give a simultaneous test statistic of all sizes in a Greig-Smith type analysis with Normally distributed "counts." His approach for a grid (based on Scheffé's S method) is complex, needs the inversion of large matrices and has no analogue for a transect. His alternative hypothesis is a "cluster," a block of size $r \times c$ of higher or lower intensity than the rest of the pattern. His simulation study shows disappointing power even against this unrealistic alternative.

6.4   ONE-DIMENSIONAL EXAMPLES

An obvious approach to the analysis of data from a transect is to regard it as a time series, remembering that the direction of "time" has no meaning. Spectral analysis has been considered by Hill (1973), Usher (1975) and Ripley (1978) for transects of counts. Unlike all the blocking methods except Hill's, it is *not* seriously affected by the starting point of the transect. In this section we illustrate the Hill, Greig-Smith, Mead, Poisson likelihood ratio test (*Plrt*) and spectral methods. The examples are taken from Ripley (1978). The values of $-2T_r$ for the *Plrt* analysis are given in Table 6.1.

The first example was a simulation of a random pattern of counts—independent Poisson observations all with mean one. The results are shown in Figure 6.4. The peaks in the Greig-Smith and Hill analyses at about block size 32 must, of course, be due to chance. (Remember the variability on those plots increases from left to right.) All the values of $-2T_r$ and Mead's test are fairly typical values for their assumed distributions. The cumulative periodogram should be within the central band with probability 0.95. The spectral density should be approximately constant. The sampling fluctuation expected at each ordinate is indicated by the double-headed arrow. The equivalent block sizes shown are one-half the wavelength.

Table 6.1   Values of $-2T_r$ for the Data of Figures 6.4–6.8[a]

Block Size	df	Figure 6.4	6.5	6.6	6.7	6.8	
1	64	79.4	62.8	100.5	35.6	199.5	(256)
2	32	32.3	82.0	117.7	32.8	200.2	(128)
4	16	17.3	68.3	122.1	55.1	129.6	(64)
8	8	7.10	6.29	121.7	56.7	80.9	(32)
16	4	6.00	6.50	5.28	26.3	77.4	(16)
32	2	5.02	4.44	7.26	24.9	59.9	(8)
64	1	0.67	1.08	0.87	8.52	21.6	(4)
128	—	—	—	—	—	2.42	(2)
256	—	—	—	—	—	44.8	(1)

[a](Except Figure 6.6, which is another simulation of the same process.) The nominal degrees of freedom given are for the first four columns; those for the last column are given in parentheses.

Figure 6.5f illustrates a simulation of a random pattern on an area with varying fertility as found in the Mercer-Hall data in Chapter 5. The mean of the Poisson count is $(1 + \sin cx)$, where the wavelength $2\pi/c$ is 10 quadrat sides. Mead's analysis gave a significant result at size 2 and a nearly significant one at size 8. Hill's analysis has a peak at block size 5, and most significantly, a trough at block size 10. These are reflected to a lesser degree in Greig-Smith's plot at sizes 4 and 8. The spectral analysis clearly picks out the periodicity in "fertility." The *Plrt* analysis gives extremely significant results at block sizes 2 and 4.

Figure 6.6f shows a pattern of clumps which was sampled by a line transect. The clump diameter is about 4.5 and the mean spacing between clumps about 10, in units of a quadrat side. Neither Hill's nor Greig-Smith's analyses pick out these features although on another simulation both showed peaks at block size 8. Mead's analysis gave a significant result at block size 2 (and at 4, 8, and 16 on the second simulation). The spectral analysis indicates a significant domination by low frequencies and has a peak at equivalent block size 10. For the *Plrt* analysis the result at block size 1 is extremely significant. We take this to indicate a generalized distribution for the counts and look at ratios of $-2T_r$. For this analysis on the second simulation block sizes 2, 4, and 8 were individually significant at 1%. The second simulation gave better results than the first, but none of the analyses clearly picks out the features in the original pattern that are obvious by eye.

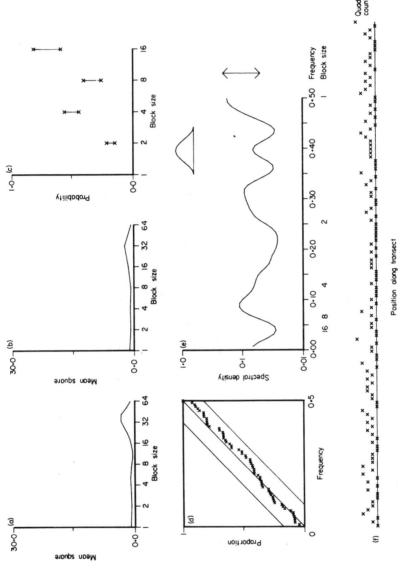

Fig. 6.4 Analyses of a random pattern of counts along a transect of 128 quadrats. The top row shows the results of the pattern analyses suggested by (*a*) Hill (*b*) Greig-Smith, and (*c*) Mead (for which high or low values of both crosses should indicate pattern at that scale). The second row shows (*d*) the cumulative periodogram with its 95% confidence band, and (*e*) an estimate of the spectral density (the smoothing window is inset and the double arrow represents the 95% confidence interval at each frequency). The counts are shown at (*f*). (Figures 6.4–6.8 are reprinted with permission from Ripley, 1978.)

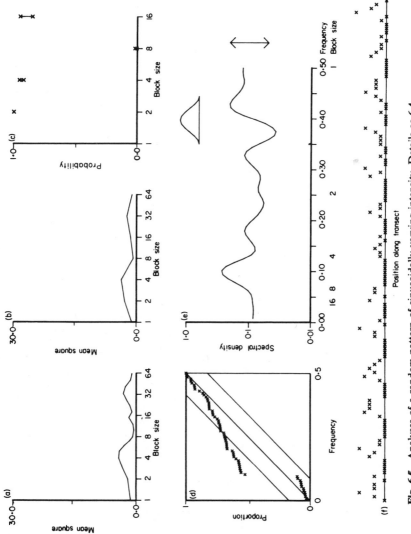

Fig. 6.5 Analyses of a random pattern of sinusoidally varying intensity. Details as 6.4.

115

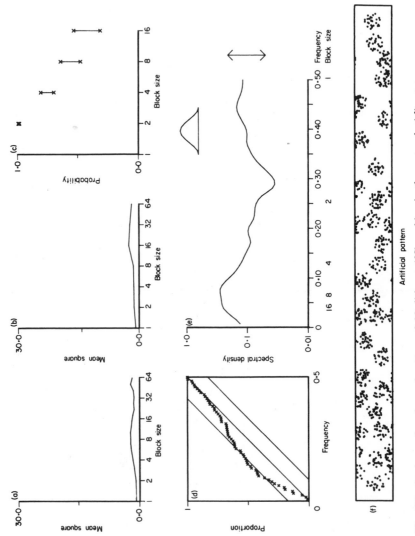

Fig. 6.6 Analyses of a clumped pattern of 30 disks at 40% packing in the rectangle (*f*). Other details as 6.4.

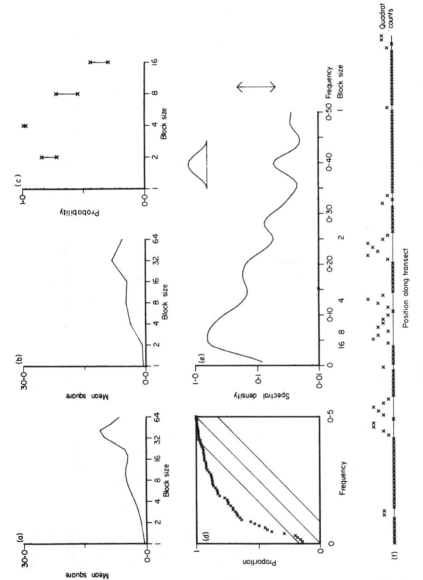

Fig. 6.7 Analyses of a transect of 128 quadrats measuring cover of mat grass. Details as in 6.4. Data from Kershaw (1957) and Mead (1974).

118

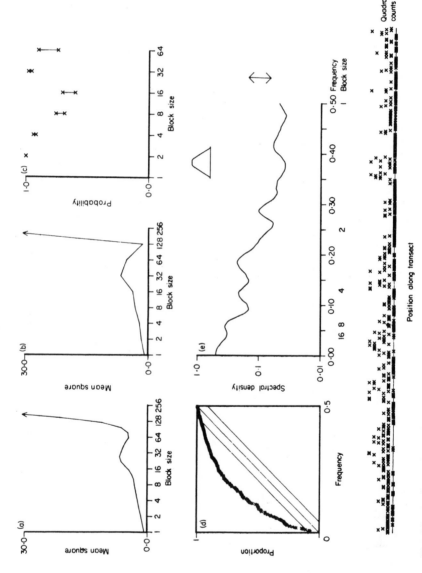

Fig. 6.8 Analyses of a transect of 512 quadrats measuring cover of *Lotus corniculatus*. Details as 6.4. Data from Usher (1975).

Figures 6.7 and 6.8 show the analyses of two sets of real data.   The first is one of eight transects taken by Kershaw (1957) and reanalyzed by Mead (1974) and Douglas (1975).   The observations were counts on a score of 0 to 5 of cover of *Nardus stricta*, a grass of moorland.   The Mead analysis indicates a high value at block size 4, but only at 10% overall significance; the analysis of all eight transects has both block sizes 2 and 4 significant at 5%.   The spectral analysis shows a dominance by low frequencies and a peak at about equivalent block size 8, which is maintained in the analysis of all eight transects.   The *Plrt* has a *low* value at block size 1, perhaps revealing the non-Poisson character of the observations.   The value at size 2 is typical of a χ^2_{32} distribution whereas sizes 4, 8, 16, 32, and 64 all give extremely high values, comparably so according to *F* tests.   This suggests significant effects at sizes 1, 2, and 4; this analysis is probably detecting the structure of the nonempty quadrats.

Our final analysis is a transect of 512 quadrats measuring cover of *Lotus corniculatus*, Bird's-foot Trefoil.   Only block size 2 of the Mead values is significant at an overall 5% level, although the value at 32 comes close. The spectral analysis suggests merely an overall decrease of power with frequency, so counts in neighboring quadrats are quite highly correlated. Both the Greig-Smith and *Plrt* analyses suggest that the two halves of the transect are quite different in character, and have particularly high values at block size 32.

These analyses are rather disappointing but spectral analysis seems the most reliable.

6.5   TWO-DIMENSIONAL EXAMPLES

We reanalyze four of the examples of Zahl (1974, 1977) to compare Greig-Smith, *Plrt* and spectral analysis with his results.   The first two examples illustrated in Figures 6.9 and 6.10 are from Thompson (1958) and are counts in 1-meter squares.   The others shown in Figures 6.11 and 6.12 are counts in 3×2-meter rectangles.   Zahl's conclusions from his own method were patches of sizes 2×2 in Figure 6.9, 6×6 in Figure 6.10, 2×2 in Figure 6.11, and about 10×10 in Figure 6.12.   All four examples are on a 16×16 grid.   The spectral analyses reveal a smooth isotropic spectral density that decreases steadily with frequency and isotropic locally positive correlations, except in Figure 6.11, which shows a marked anisotropy and a significant peak in the spectral density at a frequency $(6\pi/8, 0)$, a period of 2.7 meters in the *x* direction.

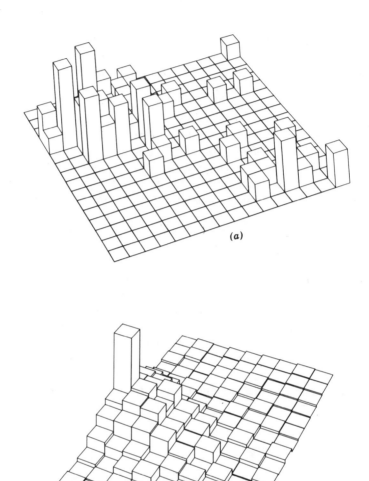

(a)

(b)

Fig. 6.9 Thompson (1958) sample A2 of bush clover. (*a*) Data. (*b*) Correlogram. (*c*) Periodogram. (*d*) Smoothed spectral density on $\log_{10}$ scale with 95% confidence interval range of 0.38. Other details as Figure 5.1.

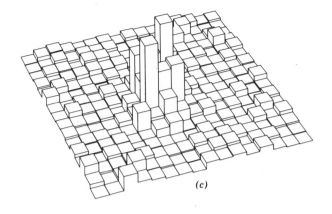

(c)

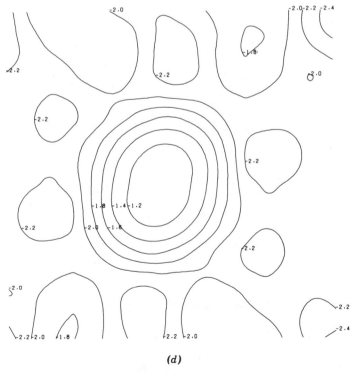

(d)

Fig. 6.9 (*continued*)

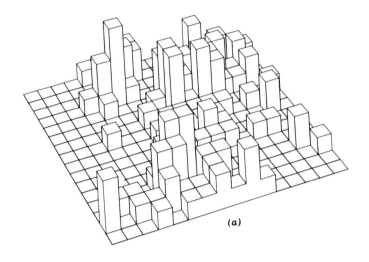

(a)

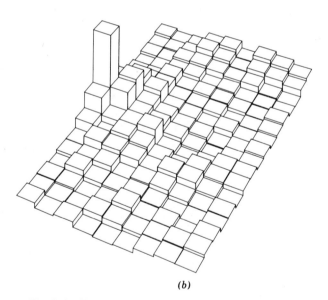

(b)

Fig. 6.10 Thompson (1958) sample A3. Details as 6.9.

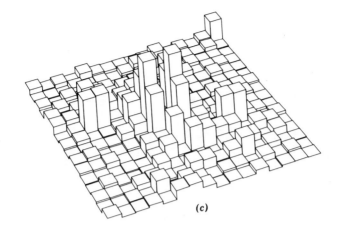

(c)

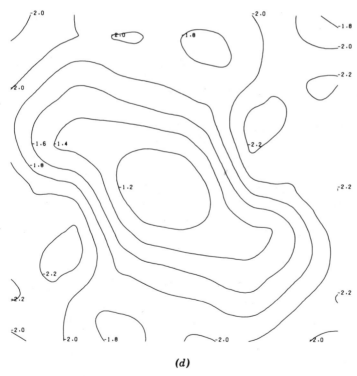

(d)

Fig. 6.10 (*continued*)

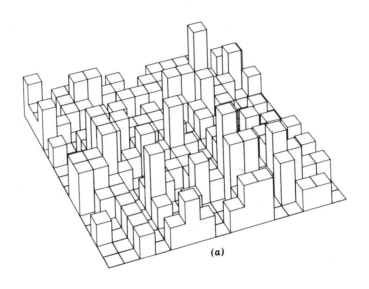

(a)

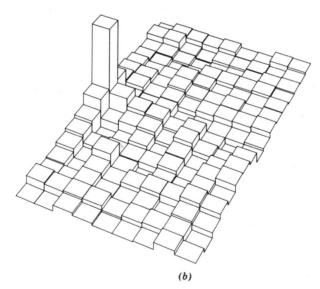

(b)

Fig. 6.11 Creosote bush sample ARIZ1 from Zahl (1974). Details as 6.9.

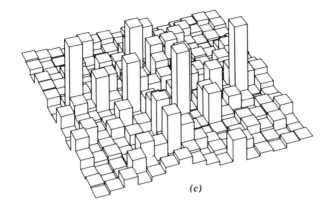

(c)

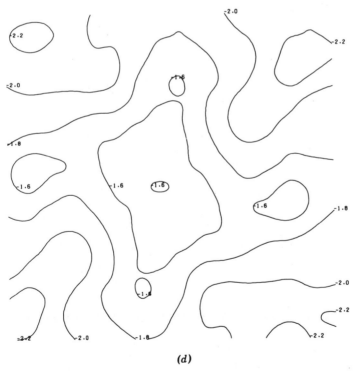

(d)

Fig. 6.11 *(continued)*

125

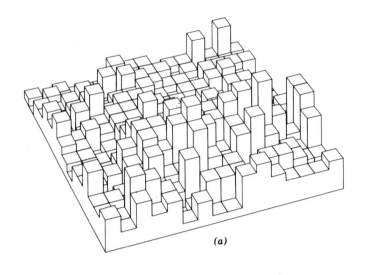

(a)

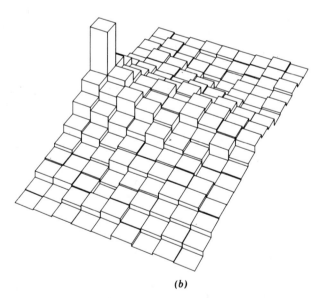

(b)

Fig. 6.12 Creosote bush sample ARIZ4 from Zahl (1974). Details as 6.9.

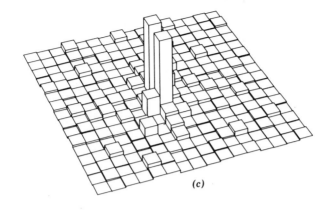

(c)

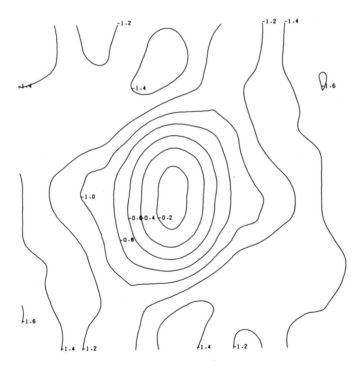

(d)

Fig. 6.12 (continued)

127

**Table 6.2 Mean Values for Square Blocks in Greig-Smith's
Analysis for the Data of Figures 6.9a–6.12a**

Block Size	Figure			
	6.9	6.10	6.11	6.12
8×8	9.10	3.77	1.27	105.0
4×4	0.77	6.45	1.47	11.2
2×2	0.61	0.72	0.78	4.39
1×1	0.50	0.64	0.66	2.73

Table 6.2 gives the mean squares for square blocks. For the first
example the large mean square at 8×8 is due to the division into quarters,
not into either 8×16 or 16×8 blocks, as should be clear from Figure 6.9a.
For the fourth example the large reduction at 8×8 is due entirely to the
division into the front and rear halves. This example shows signs of
heterogeneity both here and in the large power at low frequencies in the
spectral analysis.

Table 6.3 gives the deviances for the *Plrt* analysis, expressed as standard
Normal deviates. Both the first and third examples fit the hypothesis of a
random pattern giving independent Poisson counts with the same mean.
However, the improvement in fit when the four 8×8 blocks are allowed

Table 6.3 Values of $\sqrt{(-4T_r)} - \sqrt{(2\nu - 1)}$ for Plrt Analyses[a]

Block Size	Figure			
	6.9	6.10	6.11	6.12
16×16	1.69	4.16	1.17	8.38
8×16	1.02	4.79	1.22	8.41
16×8	0.74	4.12	1.07	4.55
8×8	-1.97	3.57	1.07	3.98
4×8	-2.02	1.60	0.99	3.60
8×4	-2.48	1.86	0.55	3.54
4×4	-2.48	-0.70	0.47	2.66
2×4	-2.74	-0.86	0.18	2.83
4×2	-3.68	-1.34	0.50	1.74

[a] ν is the number of degrees of freedom. Variables given have
asymptotically independent standard Normal distributions.

different means in 6.9 is extremely significant.   Other significant improvements in fit are 4×4 blocks in the second example, 8×4 blocks in the third and the splits into halves and quarters in the fourth.   (In this example a generalized distribution was assumed and F ratios were used.)   The periodicity found by spectral analysis is unlikely to show up in these analyses but is perhaps reflected in the significance of the 8×4 block size.   Zahl's own results are rather different.   (Note that revised conclusions for the Creosote Bush examples are given in the 1977 paper.)

We noted in Section 5.2 that Besag (1974) fitted an autologistic model to a grid of counts.

CHAPTER 7

Field Methods for
Point Patterns

The quadrat counts discussed in Chapter 6 are one class of methods which can be used to census a collection of plants or to investigate its pattern. Quadrats are reported to be difficult to use in forestry and "distance methods" have been used for decades by foresters. Their basis is the idea that if the forest is dense distances measured from a point or tree to the nearest tree will be small, so measuring such distances should allow an estimate of the number of trees per unit area to be made. However, it was found that the estimator should depend on the pattern of the trees. One line of research has been to find *robust* estimators which are little affected by the pattern. The other was to take two estimators and compare them. If they react differently to certain aspects of the pattern the comparison should give a measure of the pattern. Usually this is phrased as "testing for randomness," which means testing the null hypothesis of a forest generated by a Poisson process. The class of possible processes (which we consider in more detail in Section 8.4) is so large that a single test statistic gives insufficient information about alternative patterns. The methods of this chapter are best suited to preliminary investigations. A detailed study needs a map of the population and the methods of Chapter 8.

Distance methods have not been totally satisfactory in forestry, so Section 7.2 discusses the alternatives used. The preferred method is based on searching not for the nearest tree, but for all trees out to a fixed distance. These methods are also unsuitable for mobile animals, which are often censused from a line transect through the population. Distances are measured from the line to the animal's location when it is spotted or flushed. Section 7.3 presents the theory behind this method.

130

7.1   DISTANCE METHODS

To reduce the load on the word "point," we will suppose our point pattern to be composed of trees.   Figure 1.1a is an example.   The measurements made are of two basic types—distances from a sample point to a tree or from a tree to a tree.   Sometimes "distance methods" and "nearest neighbor methods" are used to distinguish the two types, but we will regard these terms as synonymous.   Figure 7.1 illustrates the measurements that have been suggested.   Figure 7.1a, b shows the two simplest, from a randomly chosen point and tree to the nth nearest (distinct) tree. It is natural to work with squared distances, for $\pi \times$ distance^2 is the area searched in concentric circles from the sample point or tree until the nth nearest tree is found.   To randomly select a tree it is usually thought necessary to enumerate all the trees in the study area and pick one from a table of random numbers.   This defeats the whole aim of distance methods so such measurements have not been used in practice and the last two suggestions are alternatives.   In Figure 7.1c the sample tree is taken as that nearest to the sample point.   This is *not* a tree chosen at random

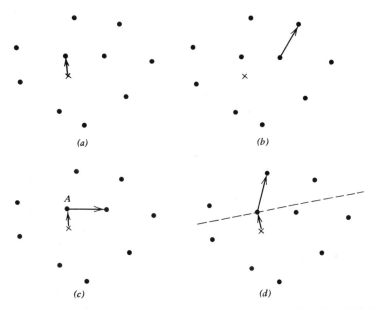

(a)   *(b)*

(c)   *(d)*

Fig. 7.1   Types of distance measurement.   ● denotes a tree, × a sample point. (a) Point to nearest tree. (b) Tree to nearest tree. (c) Tree A to nearest tree, where A is the nearest tree to the sample point. (d) T-square sampling.

Fig. 7.2   A clustered pattern. The center tree of a cluster will be chosen by the scheme of Figure 7.1c much less often than will those on the outside.

(Figure 7.2 shows that some trees will be chosen more often than others), and the two measurements are not independent. T-square sampling (Besag and Gleaves, 1973) allows a search only over a half-plane to ensure that the areas searched in the two measures cannot overlap.

Distribution Theory

The distribution theory for these measurements is simple only for a homogeneous Poisson process. Suppose we have a Poisson forest of intensity λ. If $u_n = \pi d^2$ for the distance d from any fixed point to the nth nearest tree

$$P(u_n > \pi t^2) = P(\text{no more than } n-1 \text{ trees in a ball of radius } t)$$

$$= \sum_0^{n-1} \exp\{-\lambda \pi t^2\}(\lambda \pi t^2)^r/r! \tag{7.1}$$

so u_1 has an exponential distribution rate λ. Furthermore, since the areas searched $u_1, u_2 - u_1, u_3 - u_2, \ldots$, are disjoint, these are independent exponentials each of rate λ. Hence u_n has a gamma (λ, n) distribution, as can be shown directly from (7.1). These results hold for *any* fixed sample point and hence also for a random sample point chosen independently of the process.

The same distribution theory holds asymptotically for measurements from a randomly chosen tree. Suppose we have N trees within the region D from which the tree was chosen. Let us argue conditionally on N and consider the sample tree separately; the other $N-1$ trees are independently and uniformly distributed within D (Section 2.3). Then if $A =$ area (D)

$$P(\text{no trees in an area } a | N \geqslant 1) = E\left[(1 - a/A)^{N-1} | N \geqslant 1\right]$$

$$= \{e^{-\lambda a} - e^{-\lambda A}\} / \{(1 - e^{-\lambda A})(1 - a/A)\}$$

$$(7.2)$$

We must condition on $N \geqslant 1$ to have at least one tree from which to choose. We can approximate (7.2) by $e^{-\lambda a}$, provided $a \ll A$ and $E(N) = \lambda A$ is large enough (say $\geqslant 10$). These conditions will be met in practice and lead to the same distribution theory for u_1 as before.

The distribution theory of the measurements in Figure 7.1c is not simple and has been worked out for a homogeneous Poisson forest by T. F. Cox and T. Lewis (1976). They find a function R of the two measurements, which is uniformly distributed. However, the modification to T-square sampling does have a simple distribution theory. Because the areas searched are disjoint they have independent exponential distributions with rate λ.

We noted that selecting a random tree was impracticable. Byth and Ripley (1980) introduced a scheme to select a tree with a modest amount of counting. Suppose D is divided into regions E and $F = D \backslash E$ and that the random tree is chosen from those within E. The Poisson processes on E and F are independent. By the arguments used above the probability that the disk C of radius t centered on a random tree in E does not contain further trees is

$$P(C \text{ is empty} | N(E) \geqslant 1) = e^{-\lambda a}(e^{-\lambda b} - e^{-\lambda e}) / \{(1 - e^{-\lambda e})(1 - b/e)\}$$

where $a =$ area $(C \backslash E)$, $b =$ area $(C \cap E)$, $e =$ area (E). If we have $b \ll e$, $\lambda b \ll \lambda e$, then

$$P(C \text{ is empty} | N(E) \geqslant 1) \approx \exp\{-\lambda \pi t^2\} \qquad (7.3)$$

from which we again deduce an exponential distribution for the area searched to the nearest tree. Suppose we want m samples of random tree-to-tree distances. Choose E large enough to contain $5m$ trees but divided into m widely separated small regions. Then $b = e/m$ and $\lambda(e - b) \approx \lambda e \approx 5m$; so (7.3) holds.

In deriving (7.1) and (7.3) we have assumed that we could, if necessary, measure distances from points within the study region D to trees outside D. To avoid this we usually select a random point or tree from a

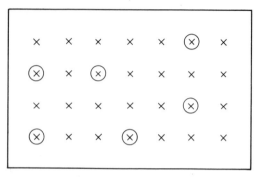

Fig. 7.3 The recommended sampling plan. Point-to-tree distances are measured from crosses, random trees chosen without replacement from those within the circles.

subregion D_0 away from the boundary of D. A further constraint is imposed if we wish to take more than one sample. To be able to regard these as independent the probability that the areas searched might have overlapped must be negligible. This is most easily achieved by taking a regular grid of sample points. The recommended sampling plan is shown in Figure 7.3. For measurements as in Figure 7.1a, c or 7.1d start from the regular grid. For both point–tree and tree–tree measurements use some grid points as sample points and the remainder to center small areas within which we enumerate the trees and then draw them at random (without replacement). A regular coverage should give the most accurate estimate of the average behavior over the region D by analogy with the theory of Chapter 3.

Intensity Estimation

Let $d_1, \ldots, d_m$ be the distances from m sample points to their nth nearest trees. For a homogeneous Poisson process, the maximum likelihood estimator of λ (based on $d_1, \ldots, d_m$) is

$$\hat{\lambda} = mn / \pi \sum d_i^2 \qquad (7.4)$$

This is not unbiased, but $\hat{\theta} = 1/\hat{\lambda}$ estimating the area per tree θ *is* unbiased. Note that $\hat{\theta}$ is just the average area searched per tree found. First moments have also been used. For $n = 1$, the estimator

$$\tilde{\lambda} = 1 / \left(2 \sum d_i / m \right)^2$$

has been proposed, because $E(d_i) = 1/2\sqrt{\lambda}$. Both $\hat{\lambda}$ and $\bar{\lambda}$ are consistent as $m \to \infty$ for a Poisson forest, but $\hat{\lambda}$ is marginally more efficient (Kendall and Moran, 1963, p. 38). Unfortunately, the constants π/n and 2 are only appropriate for a Poisson forest. Many researchers have evaluated the appropriate constants for triangular and rectangular lattices and for distances measured only in some angular sector from the sample point. Holgate (1972), Persson (1964, 1965, 1971), and Cox (1972) provide some recent surveys and comparisons. None of these estimators is particularly robust.

The other approach has been to combine estimators, taken by Diggle (1975, 1977a), Lewis (1975), Cox (1976), and Aherne and Diggle (1978). For a regular pattern distances from points to trees and from trees to trees will have, respectively, smaller and larger means than for a Poisson forest of the same intensity. The reverse is true if the trees are clustered. Clearly, some combination of intensity estimators based on the two types of measurement might be more robust than one based on just one type. Let u_i and v_i be the areas swept out by each of m point–tree and tree–tree measurements. Diggle recommended as a result of a mainly simulation study

$$\lambda^* = \sqrt{\left(\hat{\lambda}_p \hat{\lambda}_t \right)}$$

$$\hat{\lambda}_p = m \bigg/ \sum_1^m u_i, \qquad \hat{\lambda}_t = m \bigg/ \sum_1^m v_i \qquad (7.5)$$

Aherne and Diggle recommend using (7.5) if the tests of randomness discussed later reject a Poisson forest; otherwise (7.4) applied to all $2m$ squared distances (the harmonic mean of $\hat{\lambda}_p$ and $\hat{\lambda}_t$). An intuitive reason for using the geometric mean is that for clustered patterns $\hat{\lambda}_p$ tends to measure the intensity of clusters and $\hat{\lambda}_t$ the numbers within clusters, so a multiplicative correction may be appropriate. Diggle's conclusions are based mainly on biases of estimates of $\theta = 1/\lambda$. For usual values of m the sampling fluctuations of $\hat{\lambda}_p$ and $\hat{\lambda}_t$ about their means will be small, so the biases in θ and λ are closely related and good measures of robustness. The tree–tree measurements can be made by random selection of trees, by the Byth–Ripley scheme or by T-square sampling. Diggle preferred T square for its practicality. If the forest has a markedly varying intensity a systematic sampling plan will be beneficial.

Batcheler and co-workers have taken yet another approach (see the references at the end of this section).

Tests of Randomness

Many tests of the null hypothesis of a Poisson forest based on nearest-neighbor measurements have been proposed. Three recent comparative studies are given by Diggle et al. (1976), Hines and Hines (1979), and Byth and Ripley (1980). Let u_i, v_i, ut_i, vt_i be the areas swept out in point–tree, tree–tree, and in point–tree–tree T-square sampling, and $u_i^{(n)}$ the areas swept out in finding the nth nearest tree to a point. All these measurements are π times a squared distance except vt_i, for which the multiplier is $\pi/2$ (since only a semicircle is searched). Some of these tests are:

Skellam-Moore

$$n \sum_1^m u_i / \text{area}(D)$$

where D contains n points. For a Poisson forest this has a $\chi^2_{(2m)}$ distribution. Knowledge of n is unlikely and Pielou (1959; see also Mountford, 1961) suggested replacing $n/\text{area}(D)$ by a quadrat count estimate of the intensity.

Hopkins

$$Hop_F = \sum_1^m u_i \left/ \sum_1^m v_i \right.$$

$$Hop_N = \sum_1^m \{u_i / (u_i + v_i)\}$$

were proposed by Hopkins (1952) and Byth and Ripley (1980), respectively, having $F(2m, 2m)$ and $N(\frac{1}{2}, 1/12m)$ null distributions (in fact, the latter is a very good approximation to the average of m $U(0,1)$ variables). In Hop_N the pairing of u_i and v_i is arbitrary.

Holgate

$$Hol_F = \sum_1^m u_i \left/ \sum_1^m \left(u_i^{(2)} - u_i\right) \right.$$

$$Hol_N = \sum_1^m \left\{u_i / u_i^{(2)}\right\}$$

proposed by Holgate (1965a), which have the same null distributions as Hop_F and Hop_N.

T-square

$$t_F = \sum ut_i \bigg/ \sum vt_i$$

$$t_N = \sum \left\{ ut_i / (ut_i + vt_i) \right\}$$

$$t_E = 2m \sum (ut_i + vt_i) \bigg/ \left\{ \sum (\sqrt{ut_i} + \sqrt{vt_i}) \right\}^2$$

The first two were introduced by Besag and Gleaves (1973) and have the same null distributions as their *Hop* and *Hol* counterparts. t_E is a modification by Hines and Hines (1979) of the statistic of Eberhardt (1967) and can be viewed as a test of the exponential distribution of ut and vt. Hines and Hines give a table of percentage points for the null distribution of t_E.

Cox and Lewis (1976) considered the average and minimum of their statistics R_i. Hines and Hines showed that the average is very similar to t_N. Both Pollard (1971) and Diggle (1977) used Bartlett's test for homogeneity of variances, recognizing that the areas searched have $\chi^2_{(2)}/\lambda$ distributions, so testing the equality of λ for the observations provides a test for the homogeneity of a Poisson forest.

The comparison studies show that Holgate's tests are not competitive with their Hopkins and *T*-square analogues. Among *T*-square tests, t_E seems the best overall, being comparable to Hop_F for clustered alternatives, but worse than Hop_N at detecting regular forests. Hop_F and Hop_N lose power only slightly when the sampling plan of Figure 7.3 is used, which needs only a small amount of enumeration and makes them practicable.

References

Aherne and Diggle (1978), Batcheler (1971, 1973, 1975), Batcheler and Bell (1970), Batcheler and Hodder (1975), Bauersachs (1942), Besag and Gleaves (1973), Brown and Holgate (1974), Byth and Ripley (1980), Campbell and Clarke (1971), Catana (1963), Clark and Evans (1954, 1955), Cormack (1977), Cottam (1947), Cottam and Curtis (1949, 1956), Cottam et al. (1957), F. Cox (1972), T. F. Cox (1976, 1979), Cox and Lewis (1976), Craig (1953), Dacey (1962, 1963), Dacey and Tung (1962), Diggle (1975, 1977a, b), Diggle et al. (1976), Eberhardt (1967), Hines and Hines (1979), Holgate (1964, 1965a, b, c, 1972), Hopkins (1954), Keuls et al. (1963), Lewis (1975), Mark and Esler (1970), Mawson (1968), Moore (1954), Morisita (1954, 1957), Mountford (1961), Ord (1978), Persson (1964, 1965, 1971, 1972), Pielou (1959, 1962), Pollard (1971), Skellam (1952), Thompson (1956).

7.2 FORESTRY ESTIMATORS

The tessellations and triangulations introduced in Section 4.3 can be used to give estimates of the intensity λ of a forest.   For the Dirichlet cells the average area is A/N for N trees in a region D of area A, simply because the N cells partition D.   If we ignore edge effects this is also true of the about $2N$ Delaunay triangles (Fraser and van den Driessche, 1972).   Thus if we could select a random cell or triangle and measure its area we would have an unbiased estimator of the area per tree.   All we can do, however, is to select a random point within D and measure the area of the cell (or triangle) containing that point.   Let the areas be $a_1, \ldots, a_N$.   Then the probability of selecting the ith cell is a_i/A, and the expectation of $1/a_i$ is $\Sigma(1/a_i)a_i/A = N/A$.   Thus $1/\text{area}$ for a cell selected in this way is an unbiased estimator of λ for any homogeneous pattern.   For cells this idea is due to Ord (1978).   Finding the Dirichlet cell or Delaunay triangle containing a point *is* a local procedure.   (The nearest tree defines the Dirichlet cell and is one of the vertices of the Delaunay triangle.)   However, constructing them in the field and measuring their area is clearly difficult.

Foresters have found laying out quadrats and actually counting accurately difficult in dense forests.   Distance methods provide insufficiently robust estimators and the ideas (proposed by foresters) described at the beginning of this section are too intricate.   They are usually most interested in the volume of timber in a tract of forest and use the Bitterlich (1948) method.   This is discussed by Grosenbaugh (1952a,b), Shanks (1954), Kendall and Moran (1967, p. 47), Holgate (1967), and Ord (1978). For trees of circular cross section and equal diameter ρ it essentially uses a circular quadrat and so gives an unbiased estimate.   The observer has an instrument known as a *relascope*, which enables him to decide whether the angle subtended by a tree trunk is greater than 2α.   From Figure 7.4 we see that this is so if the tree is nearer than $\rho/(2\sin\alpha)$, so the observer counts those trees in a circle of this radius, and estimates λ by *number* $\times$ $4\sin^2\alpha/\pi\rho^2$.   If the trees are of different or noncircular cross sections their apparent diameters ρ_i can be measured and

$$\sum_1^n 4\sin^2\alpha/\pi\rho_i^2$$

formed.   In either case, $4n\sin^2\alpha$ is an unbiased estimate of the *basal area* per unit area of forest, and when multiplied by the area of the tract and an average useable height of a trunk gives an estimate of the timber volume.

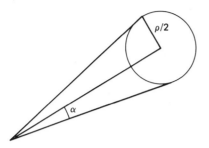

Fig. 7.4 The geometry of Bitterlich sampling.

Obviously, α should be taken as small as possible to include many trees. However, nearby trees might obscure distant trees and Holgate points out that errors in α become increasingly important at small α. A compromise of $\alpha \approx 2°$ seems satisfactory. Of course, the variance of the estimate depends on the pattern. In practice, the counting will be repeated at several observation points and the sample variance can be used to estimate the sampling errors.

7.3   LINE TRANSECTS

Mobile animal populations provide rather different problems of censusing to those of forests. Two common examples are counting deer and game birds, both of which will usually flee as an observer approaches. This flight mechanism is used in the line-transect method. An observer walks a long transect through the area to be censused and whenever an animal is detected (usually by being flushed) he marks the spot and measures either the distance d from his present position or the right-angle distance x to the animal from the transect line (see Figure 7.5). During the course of a transect of length L he spots n animals and records $(x_1, \ldots, x_n)$ or $(d_1, \ldots, d_n)$. If all animals were detected out to distance W and those beyond ignored, we would have a *strip transect* and the obvious estimator of λ, the number of animals per unit area, is

$$\hat{\lambda} = n/2LW \qquad (7.6)$$

This is effectively a quadrat estimator, using a $L \times 2W$ rectangular area. The line transect methods find an equivalent of the half-width W. Many

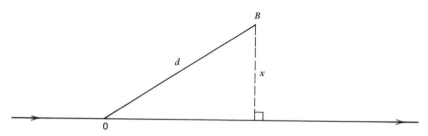

Fig. 7.5 Measurements made in line transect sampling. The observer spots an animal at B when at 0.

slightly different sets of assumptions have been proposed. We certainly need:

1. Animals only move after detection.

2. The detections of animals are independent and no animal is detected more than once.

3. The detection probability is a function $g(x)$ of the right-angle distance x from the transect, and $g(0)=1$.

Unless detection of animals on the transect is sure we will never know what the missed proportion of animals is.

Some other assumption on the pattern of the animals is required. They can be assumed to follow a Poisson or binomial process or the pattern can be considered fixed and the transect positioning random.

Suppose for the moment that $g(x)=0$ for $x>W$. Then if X is a right-angle distance

$$P(X \leqslant x| \text{ an animal is observed})$$

$$= P(X \leqslant x, \text{ animal observed})/P(\text{animal observed})$$

$$= \left\{ \int_0^x g(y)\,dy/W \right\} \Big/ \left\{ \int_0^\infty g(y)\,dy/W \right\}$$

Thus the appropriate *pdf* of X when animals are uniformly distributed in a strip width $2W$ is

$$f(x)=g(x)/\mu \qquad \mu=\int_0^\infty g(y)\,dy \tag{7.7}$$

We may now let $W\to\infty$. Then from (7.7) the observed right-angle distances $(X_1,\ldots,X_n)$ are independent with *pdf* $g(x)/\mu$, conditional on n.

However, for a binomial or Poisson process

$$E(n) = 2\lambda L\mu = 2\lambda L/f(0)$$

so

$$\hat{\lambda} = nf(0)/2L \tag{7.8}$$

is an unbiased estimator of λ. Of course, (7.6) is just the special case $g(x) = 1$ for $0 \leqslant x \leqslant W$, $g(x) = 0$ for $x > W$.

Note that g differs from a *pdf* by the scale factor μ which we must determine, and that to use (7.8) we must estimate $f(0)$ for a *pdf* f, from which we have n observations $(x_1, \ldots, x_n)$. Both parametric and nonparametric methods have been proposed to estimate $f(0)$. Suppose that we have an approximately unbiased estimator $\hat{f}(0)$ and $n\,\mathrm{var}(\hat{f}(0)|n) \approx \sigma^2$. Then for a Poisson pattern

$$E(\hat{\lambda}) \approx \lambda$$

$$\mathrm{var}(\hat{\lambda}) \approx \left[f(0)\,\mathrm{var}(n) + E\{ n^2 \mathrm{var}(\hat{f}(0)|n) \} \right] / (2L)^2$$

$$\approx \lambda(1 + \mu^2\sigma^2)/2\mu L$$

Such theoretical variance formulas depend rather heavily on the assumptions, so the variability of an estimator of λ is probably better assessed by the sample variance of estimators from different transects.

For parametric estimation it is common to take either $g(x)$ as the tail area for a family of *pdfs* or $f(x)$ as one of a family of *pdfs*. Forms used are:

$$g(x) = e^{-\gamma x}, \ f(x) = \gamma e^{-\gamma x} \qquad \text{Gates et al. (1968)}$$

$$g(x) = \begin{cases} 1 - (x/W)^\alpha & 0 \leqslant x \leqslant W \\ 0 & x > W \end{cases} \qquad \text{Eberhardt (1968)}$$

$$\left. \begin{array}{l} g(x) = \exp\{ -(x/\beta)^\alpha \} \\ \text{Weibull tail area} \end{array} \right\} \qquad \left\{ \begin{array}{l} \text{Pollock (1978)} \\ \text{Ramsey (1979)} \end{array} \right.$$

$$g(x) = \int_x^\infty \beta^\alpha y^{\alpha-1} e^{-\beta y}\, dy \qquad \text{Sen et al. (1974, 1978)}$$

$$g(x) = \exp\{ -x^2/2\sigma^2 \}$$

$$f(x) = \sqrt{(2/\pi\sigma^2)}\exp\{ -x^2/2\sigma^2 \}$$

Except for the exponential g and half-normal f, the corresponding families for f are not quite the usual probability densities and methods of maximum likelihood estimation have been developed in the papers cited. These yield estimators of $f(0)$, which are then used in (7.8). One of the simplest nonparametric methods yielding the "Kelker Index" is to suppose on biological grounds that $g(x) = 1$ for $0 \leqslant x \leqslant w$ and use (7.6) with W replaced by w and n the number of distances measured that are less than w. A "jacknifed" form is to use

$$\hat{\lambda} = (3n_1 - n_2)/(4Lw)$$

where n_1 and n_2 are the numbers of distances in $(0, w)$ and $(w, 2w)$ respectively (Eberhardt, 1978). Anderson and Pospahala (1970) suggested a local quadratic form for f; fit

$$f(x) = b_0 + b_2 x^2$$

and use $\hat{f}(0) = \hat{b}_0$.

Burnham and Anderson (1976), whose general approach we have followed in deriving (7.8), suggested

$$\hat{f}(0) = \max_i \left\{ \frac{i}{nx_{(i)}} \right\} \qquad x_{(1)} \leqslant x_{(2)} \leqslant \cdots \leqslant x_{(n)}$$

as robust to animals moving away from the transect before detection. Their main intention seems to be that modern probability density estimation techniques should be used to give a nonparametric estimate of f but these do not seem yet to have been tested.

So far we have assumed that right-angle distances are available. For radial distances d we consider $g(x, d)$ as the probability of detection at right angle distance x and radial distance d. Then

$$f(x) = E\{f(x|D)\}, \qquad 1/\mu = f(0)$$

where $f(x|d)$ is the conditional *pdf* of x given d, and we can estimate $f(0)$ by the obvious estimate, suggested by Burnham and Anderson (1976),

$$\hat{f}(0) = \frac{1}{n} \sum_1^n f(0|d_i) \qquad\qquad (7.9)$$

if the form of $f(x|d)$ is known or can be estimated. The estimator of

Hayne (1949) assumes a circular flushing area for each animal, so

$$f(x|d) = 1/d \qquad x \leqslant d, \qquad 0 \qquad x > d$$

and

$$\hat{\lambda} = \left(\sum_{1}^{n} 1/d_i \right) / 2L \quad = n/2Lh$$

where h is the harmonic mean of the d_i. Gates (1969) assumed $g(x, d) = e^{-\gamma d}$, $0 \leqslant x \leqslant d$, which would again give Hayne's estimator (since each animal has a circular flushing area of exponentially distributed radius). However, maximum likelihood estimation and some bias correction led Gates to

$$\hat{\lambda} = (2n - 1)/2L\bar{d}$$

Thus the "obvious" form (7.9) need not be the best.

The comparative studies available for this wealth of estimators are not satisfactory. Most are not comprehensive and have concentrated on a narrow class of mathematical models. Robinette et al. (1954, 1956, 1974) have constructed field trials in which Anderson and Pospahala's and Kelker's nonparametric methods were preferred to parametric methods. It seems that until the recent introduction of the Weibull and gamma tail areas, parametric methods have been insufficiently flexible.

References

Gates (1980) has an extensive bibliography of applications.
Anderson and Pospahala (1970), Burnham and Anderson (1976), Eberhardt (1968, 1978, 1979), Emlen (1971), Gates (1969, 1980), Gates et al. (1968), Hayne (1949), Kovner and Patil (1974), McIntyre (1953), Moore (1955), Pollock (1978), Ramsey (1979), Robinette et al. (1954, 1956, 1974), Sen et al. (1974, 1978), Skellam (1958b), Smith (1979).

CHAPTER 8

Mapped Point Patterns

Maps of small objects that may be considered as points are common; we may think of towns, forts, and stores in geography and archeology, plants, and nests in ecology, galaxies in astronomy, and many other examples. Throughout this chapter we will assume that our raw data are a map (or a list of coordinates) giving the locations of all N objects within a region D of known area A. Typically N will be less than 100. Thus it is essential to use all the available information efficiently. There are examples in which it is difficult to define the study region; Brown (1975) and Newton et al. (1977) discuss the locations of birds' nests that can only occur in a particular habitat within the region (this habitat had not been accurately mapped).

The assumptions of homogeneity and isotropy have only been in the background in Chapters 6 and 7, where we estimated the intensity averaged over D, or tested for departures from a homogeneous Poisson process. In this chapter we will always assume homogeneity and usually isotropy. For the continuously distributed observations of Chapters 4 and 5 we could achieve stationarity by transformations or by removing a deterministic trend. For point processes our sole recourse is to analyze small areas intensively that we assume to be homogeneous when looking for interactions between objects, and to use large scale analyses such as those in Section 6.5 to detect intensity variations within our study region.

There is a fundamental problem here. We shall see that there are mathematical models for the ideas of a Poisson process with intensity varying from place to place, and of *cluster* processes in which daughter objects are distributed about parent objects, which have identical distributions and so cannot be distinguished by any amount of data. The observer must bear in mind that what is considered to be a clustered pattern with the assumption of homogeneity in force could also be the result of heterogeneity.

Figure 8.1 illustrates a broad classification of point patterns. All are computer simulations of the processes detailed in the caption. Figure 8.1*a*

144

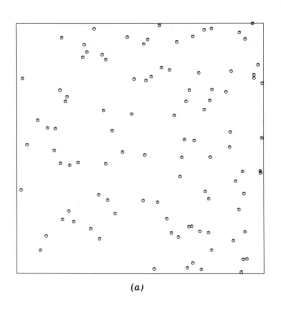

(a)

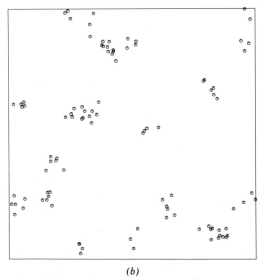

(b)

Fig. 8.1 Simulations of 100 points from (*a*) binomial process (*b*) cluster process with 20 parents and cluster radius 0.05 (*c*) cluster process with 10 parents and cluster radius 0.25 (*d*) heterogeneous binomial process (*e*) randomly packed centers of disks of diameter 0.06 (*f*) randomly packed centers of ellipses with horizontal major axis 0.06, vertical minor axis 0.04.

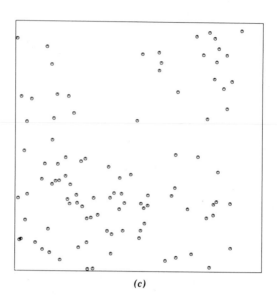

(c)

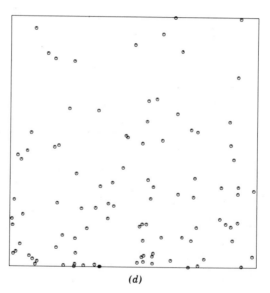

(d)

Fig. 8.1   (*continued*)

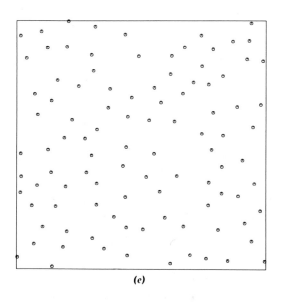

(e)

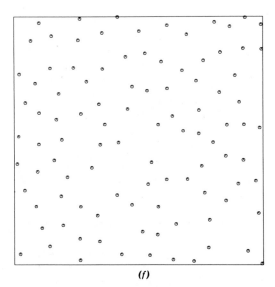

(f)

Fig. 8.1   (*continued*)

is a realization of a Poisson process. Remember from Section 2.3 that conditionally on the number of points they are independent and uniformly distributed. Most people feel that a "random" pattern should not be so chaotic with such close pairs of points and large empty spaces. Both Figure 8.1*b* and 8.1*c* are from cluster processes with (on average) 20 and 10 clusters, respectively. In the latter, the clusters are so large that we would probably regard the pattern as heterogeneous. Figure 8.1*d* is a more clear-cut case of heterogeneity with a clear vertical trend in intensity in a Poisson process. Finally, Figure 8.1*e, f* represents realizations of processes with inhibitions between the objects, which we will call *regular* patterns. The difference between them is that Figure 8.1*f* has a subtle anisotropy in the interactions and so has different properties in the horizontal and vertical directions (try turning the page 90°). The human eye (plus brain) is quite adept at classifying point patterns but is easily fooled. (Anyone who does not believe this should look at the examples on pp. 5–8 and p. 32 of Hodder and Orton, 1976.)

There are several distinct approaches to the analysis of mapped point patterns. A traditional view has been to sample by quadrats, represented, for example, by Rogers (1974), covered in Chapter 6. Perhaps the commonest approach is to use tests of "randomness" based on nearest-neighbor distances such as the test of Section 7.1. These are often misapplied, for to measure the distance from *every* object to its nearest-neighbor violates the sampling assumptions made there. In Section 8.2 we shall discuss what can be salvaged for mapped point patterns.

To measure all nearest-neighbor distances we will have to measure all small distances between pairs of objects, so it seems sensible to use all this information. This leads us to methods based on the distribution of the distances between pairs of objects. Somewhat surprisingly, this is closely related to the second-moment properties of the counts $N(A)$ of the number of points in a set A within D. We can always find $N(A)$ by adding the contributions from many subsets of A of small area. For small enough A and B, the noncentered covariance

$$E(N(A)N(B)) \approx P(N(A) > 0, \quad N(B) > 0)$$

$$\approx g(\mathbf{x}, \mathbf{y}) \times \text{area}(A) \times \text{area}(B) \qquad (8.1)$$

for points $\mathbf{x} \in A$, $\mathbf{y} \in B$. For a homogeneous, isotropic process $g(\mathbf{x}, \mathbf{y})$ must be a function of the distance $d(\mathbf{x}, \mathbf{y})$, and can be thought of as a density of the interpoint distance distribution. This approach is taken further in Section 8.3. An alternative view of the second moments is to estimate the coefficient of variation $\text{var}(N(A))/E(N(A))$ for a square A. This is akin to the David–Moore index we discussed for quadrat counts. With a mapped pattern it is possible to average analytically over all positions of the quadrat A and so to reduce the sampling variability.

Measurements on the Dirichlet cells and Delaunay triangles (see Section 4.3) defined by the pattern have been advocated by a number of authors [Boots (1974, 1975), Getis and Boots (1978), Mardia et al. (1978), Mollison and the reply in Ripley (1977), Vincent et al. (1976a) and Smalley (1966)]. Miles (1970) summarizes the known moments results for a Poisson process. The suggestions include the variance of the area of the Dirichlet cells and angle measurements on Delaunay triangles. There are no published comparative studies, but unpublished work by P. J. Green (abstract: *Advances in Applied Probability* **10**, p. 493, 1978) showed these ideas to be good, but not the best available.

With mapped data we can do much more than just test for "randomness." In Section 8.4 we consider mathematical models that formalize the ideas of clustering, inhibition, and heterogeneity, which have been in the background of Chapters 6 and 7. The estimation of parameters of these models is a largely unsolved problem. Some general ideas are presented in Section 8.4 and applied in Section 8.6 to several examples of complete analyses of point patterns.

8.1   BASIC PARAMETERS

Suppose we have observed N objects $x_1, \ldots, x_N$ within a "reasonably compact" domain D. For definiteness we will work in the plane unless otherwise stated. Several mathematical descriptions of these observations are possible and these suggest various analyses. We could use the coordinates of the object, but any analysis would have to ignore the order in which the objects are labelled and respect the supposed invariance of the process under rigid motions of the plane given by homogeneity and isotropy. Multidimensional scaling considers the reduction of information to the set $\{d_{ij}\}$ of distances between the objects; from this it is possible to recover the locations of the objects up to a rigid motion (Torgerson, 1958). We could further discard the labels that tell us which distance is associated with which pair of objects. This reduces our information to the interpoint distance distribution, which as we saw in the chapter introduction has been used as a summary statistic (Bartlett, 1964). A further way to condense $\{d_{ij}\}$ is to consider $\{d_i = \min_j d_{ij}\}$, the set of distances from each object to its nearest neighbor. This approach is taken up in Section 8.2.

The most theoretically useful approach to point processes has been via the description by $N(A)$, for all bounded Borel sets A. The reader is referred to the survey papers of Daley and Vere-Jones (1972) and Fisher (1972) and to the advanced monographs of Matthes et al. (1978) and Kallenberg (1976) for the full power of this approach. Carter and Prenter (1972) and Ripley (1976b) show the formal equivalence of all the approaches sketched here.

From the counts the natural parameters to consider are the moments. By homogeneity $E(N(A))=\lambda v(A)$, where λ is a constant, the *intensity*, and $v(A)$ the area (in $\mathbb{R}^2$) or volume (in $\mathbb{R}^3$) of the set A. The reduction of the second moments by homogeneity and isotropy is more difficult; less mathematical readers may wish to skip the rest of this paragraph. The noncentered covariances $E(N(A)N(B))$ can be reduced to a nonnegative increasing function K on $(0,\infty)$ by

$$E(N(A)N(B))=\lambda v(A\cap B)+\lambda^2 \int_0^\infty v_t(A\times B)\,dK(t) \qquad (8.2)$$

where

$$v_t(A\times B)=\int_A \sigma_t\big[\,\{y-x|y\in B, d(x,y)=t\}\,\big]\,dv(x)$$

and σ_t is the uniform probability on the surface of the sphere of radius t centered at the origin (Ripley, 1976a). The following remarks may help give the intuitive content of (8.2). The first term comes from objects counted in both A and B, the second from distinct pairs of points $x\in A$ and $y\in B$. By the assumed homogeneity and isotropy the only relevant characteristics of (x,y) are those invariant under rigid motions, hence functions of the distance $d(x,y)$. The measure v_t formalizes the idea of one object uniformly distributed in the space with the second uniformly distributed distance t from the first.

Certain special cases give intuitive interpretations of K:

1. $\lambda^2 K(t)$ is the expected number of (ordered) pairs of distinct points not more than distance t apart with the first point in a set of unit area.

2. $\lambda K(t)$ is the expected number of further points within t of an arbitrary point of the process (see Ripley, 1977, p. 190).

3. From (8.1) $g(d(x,y))$ is given by $g(t)=(\lambda^2\,dK/dt)/c(t)$, where $c(t)$ is the length or area of the surface of a ball of radius t.

For a Poisson process we find $K(t)$ is the area or volume of a ball of radius t (πt^2 or $4\pi t^3/3$) as we might expect from 1, 2, or 3. The power of λ in the definition of $K(t)$ was chosen to give this result.

The higher moments of $\{N(A)\}$ reduce in a similar way, but to unwieldy functions. There is some evidence (Julesz, 1975) that the human eye is most aware of the second order properties of planar point patterns.

Yet another description of a point pattern is obtained by retaining only whether A contains an object, that is replacing $N(A)$ by $\min(N(A),1) = I(N(A)>0)$. This point of view will be extended to random sets in Section 9.1. We do not need to consider all Borel sets A; the distribution of the stochastic process $\{N(A)\}$ is specified by giving

$$P(N(A_1)=0,\ldots,N(A_r)=0)$$

for all finite collections $(A_1,\ldots,A_r)$ of balls (Ripley, 1976b). The "first-order" part of this description would be

$$p(t)=P(N(\text{ball center } \mathbf{x}, \text{radius } t)>0) \tag{8.3}$$

which by homogeneity does not depend on $\mathbf{x}$. However, $p(t)$ is the cumulative distribution function of the distance from $\mathbf{x}$ to the nearest object. Cowie (1967), Roder (1975), and Diggle (1979) have suggested considering the analogous distribution for object–nearest-object distances.

Gaussian processes are characterized by their mean and covariance functions. No combination of the parameters presented in this section characterizes an interesting class of point processes, yet in suitable combinations they can provide insights into which mechanisms could and could not have generated an observed point pattern, as we shall see in Section 8.6.

Edge Effects

Interpretation (1) of K shows that it is closely related to the distribution F of the distances of pairs of points *within* D. The latter depends strongly on the shape and size of D. For a (homogeneous) Poisson process it is the distribution of the distance between a pair of independent uniformly distributed points in D, formulas for which were given in specific cases by Borel (1925) and for domains with smooth boundaries (an unstated condition) by Geciauskas (1977).

The effect of the edge of the domain D becomes increasingly dominant as the dimension increases. A number of special edge corrections are considered for individual methods. In addition, there are two general approaches:

1. Consider D within some larger domain D^* and allow measurements from objects in D to objects in D^*. Usually this can be achieved only by

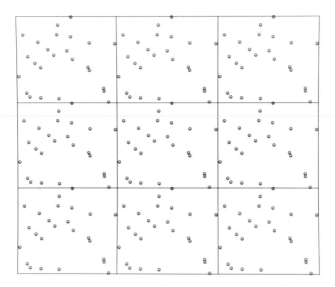

Fig. 8.2 An interpretation of toroidal edge correction. Distances may be measured from a point in the central rectangle to points in the surrounding rectangles. (From Ripley, 1979b.)

defining a new study region inside the map as given. The effective sample size is then the number of points in D^*. This method is usually referred to as allowing a *border* or *guard area*.

2. Rectangular regions D can be regarded as a torus, so that points on opposite edges are considered to be close. An alternative interpretation illustrated in Figure 8.2 is to regard D as part of a grid of identical rectangles, forming a border with copies of the pattern inside D.

8.2 NEAREST-NEIGHBOR METHODS

In Chapter 7 we mentioned the Skellam–Moore test; $d^* = N(\pi\Sigma_1^N d_i^2)/A$ is referred to a $\chi^2_{(2N)}$ distribution to test the null hypothesis of a Poisson process. This is, however, a misuse of the test, for the derivation of the χ^2 distribution assumed that the squared distances d_i^2 and d_j^2 are independent. However, there is a nonnegligible probability that the areas searched for the nearest neighbors will overlap; Donnelly (1978b) has shown by numerical integration that the correlation between d_i^2 and d_j^2 is enough to reduce the variance of d^* by about 16% from that predicted by the χ^2 distribution (in two dimensions, using toroidal edge correction, for all N).

Brown and Rothery (1978) consider this statistic and two others introduced by Brown (1975)—the coefficient of variation S and the ratio G of the geometric mean to the arithmetic mean of $\{d_i^2\}$. Brown needed a test for regularity of birds' nests which did not depend on the area A of D, which was only vaguely specified. Both S and G have this property.

Dependence is only one of the problems in applying these tests to maps. We also have to counter edge effects. Since we interpret πd_i^2 as the "area searched" from the ith object we could substitute the actual area searched within D. Even with this edge correction, no adequate approximation to the sampling distribution of d^*, S, or G is known, so only Monte Carlo tests (Section 2.5) are available.

Clark and Evans (1954) advocated the average of the distances (d_i) rather than that of their squares. Let $\bar{d}$ be this average. Then they introduced $R = \bar{d}/E(d_i)$ as a "measure of nonrandomness," large for regular patterns and small for clustered or heterogeneous data, and as a test statistic,

$$CE = \frac{\left[\bar{d} - E(d_i)\right]}{\sqrt{\operatorname{var}(\bar{d})}} \tag{8.4}$$

to be referred to the standard Normal distribution (although because of dependence this does *not* follow from the usual Central Limit Theorem). The expectation and variance are for a binomial process with N points. Ignoring both edge effects and correlations between d_i and d_j, we find

$$E(d_i) \simeq 0.5\sqrt{(A/N)}$$

$$\operatorname{var}(\bar{d}) \simeq (4-\pi)A/4\pi N^2 \simeq 0.0683 A/N^2$$

from the moments of an exponential distribution, substituting $\hat{\lambda} = N/A$ for λ. Diggle (1976) and Donnelly (1978a) investigated the correlation between d_i and d_j and found it to be small. Clark and Evans recommended a border to combat edge effects but many users of their test have since ignored that advice. The problem has been investigated by Matérn (1972), Persson (1972), De Vos (1973), Hsu and Mason (1974), Vincent et al. (1976b), and Donnelly (1978a, b), with various remedies. Donnelly showed that the distribution of $\bar{d}$ is Normal for all practical purposes for N greater than six, so all that is needed are approximations to its mean and variance. He found, by a mixture of numerical integration and simulation,

$$E(d_i) \simeq 0.5\sqrt{(A/N)} + (0.514 + 0.412/\sqrt{N})P/N$$

$$\operatorname{var}(\bar{d}) \simeq 0.070 A/N^2 + 0.037 P\sqrt{A}/N^{5/2} \tag{8.5}$$

for a rectangle D of area A and perimeter length P. (If toroidal edge correction is used, of course $P = 0$.) Note that the effect of ignoring these edge corrections is to bias the test, so that regularity will be detected spuriously. The use of mean distances rather than mean squared distances has been criticized as inefficient, for squared distances give the maximum likelihood estimator of λ for a Poisson process. For a test this is irrelevant; d^* and $\bar{d}$ can only be compared in simulation studies such as those reported in Section 8.5.

At (8.3) we define $p(t)$ as the cumulative distribution function of the distance from a point to the nearest object. The obvious way to estimate this for a homogeneous pattern is to find the empirical distribution function for distances from m sample points to their nearest objects. However, edge effects intervene, so for each sample point we measure the distance d_i to the nearest object and r_i to the boundary and compute

$$\hat{p}(t) = \#\{i \mid d_i \leqslant t \leqslant r_i\} / \#\{i \mid t \leqslant r_i\}$$

Thus for each t we only consider the subsample of points at least t from the boundary (reply to Ripley, 1977). How should these m points be chosen? From the point of view of Chapter 3 we are sampling to find the average of a random function defined on D, the distance from a point to its nearest object. We would expect large positive local correlations, so the theory suggests systematic or stratified random sampling. Since a general-purpose program *might* meet a completely regular pattern it is probably best to use stratified random sampling.

Baddeley (1980) has shown weak convergence of $\hat{p}$ to p at rate $1/\sqrt{N}$ for a class of point processes but unfortunately the limit process is not a standard process even for a Poisson point process. Since the sampling variability will be proportional to $1/\sqrt{m}$ this suggests we take m proportional to N, at least for small N. For large N we will probably decide on a particular accuracy for $\hat{p}$ and choose m accordingly, irrespective of N. Figure 8.3 shows the results of a simulation experiment. First, 10 realizations of a binomial process of 50 points were sampled by a 50×50 stratified random sample; then, a single realization was sampled 10 times each by 10×10, 30×30, and 50×50 samples. The results suggest that m in the range $10N$ to $20N$ with a maximum of $m = 1000$ is sufficient. (Ripley, 1979c, recommended $10N$, whereas Diggle, 1979, used 10×10 for $N = 100$, which seems insufficient. cf. Diggle and Matérn, 1980.)

Bartlett (1974, 1975) looked at histograms of the distribution of $\{d_i\}$ object–nearest-object distances rather than point–nearest-object distances. He computed the Eberhardt (1967) index $E = N \sum d_i^2 / (\sum d_i)^2$, the sampling

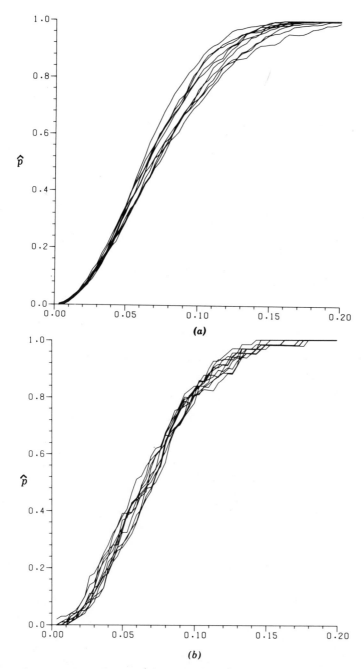

Fig. 8.3 $\hat{p}$ for a binomial process of 50 points. (*a*) 10 different realizations, 50×50 grid (*b*), (*c*), (*d*) 10 samples of the same realization, 10×10, 30×30 and 50×50 grids.

155

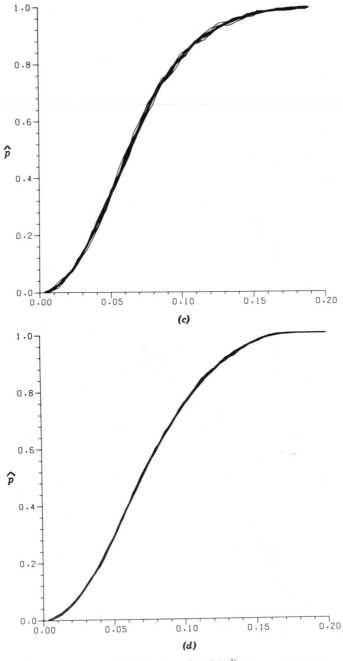

Fig. 8.3 (*continued*)

theory of which seems never to have been investigated. Diggle (1979) considered the empirical cumulative distribution function $\hat{q}$ of $\{d_i\}$ (previously suggested by Cowie, 1967 and Roder, 1975) and defined the test statistics

$$d_x = \sup|\hat{p}(t) - p(t)|$$

$$d_y = \sup|\hat{q}(t) - p(t)|$$

$$d_2 = \sup|\hat{p}(t) - \hat{q}(t)|$$

where $p(t) = 1 - \exp(-\pi N t^2)$ is the approximate mean of $\hat{p}(t)$ and $\hat{q}(t)$ for a binomial process of N points. The use of a Kolmogorov-type test is not particularly appropriate, as the standard error of $\hat{p}(t)$ will vary with t as illustrated in Figure 8.3. Only Monte Carlo tests have been used, although Baddeley's asymptotic theory suggests that $\sqrt{N}d_x$ will have a limit distribution and a table of percentage points could be created. Diggle's experiments suggest that d_x is effective against clustering (or heterogeneity, for which $\hat{p}$ was originally introduced) and d_y against regularity, d_2 being a compromise between the two.

Note that $\hat{p}(t)$ could be computed analytically within each Dirichlet cell (for which the defining object is the nearest) and the results for each cell averaged, weighted by the cell area. The computation involved is heavy, but knowledge of the Dirichlet cell contiguities can be used to speed up the search for nearest neighbors. Suppose that the stratified sampling grid of points is scanned as in Figure 8.4. Then the nearest object is usually either the nearest object for the previous point or the object defining one of its contiguous cells. Further, this can be confirmed by checking if the point falls within one of these cells. Finding the contiguity information is an appreciable overhead, however, as the following timings show, for

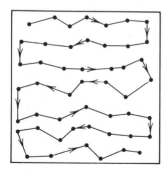

Fig. 8.4 Order of scanning of a grid for $\hat{p}$.

$N = 50$, $m = 500$:

Naive evaluation	2.4 sec	
Contiguity information	0.32 sec	values for CDC 6400
Computing distances	0.46 sec	

A nearly systematic sampling scheme brings computational as well as statistical benefits!

References

Atkinson (1974), Baddeley (1980), Blackith (1958), Blackith et al. (1963), Brown (1975), Brown and Rothery (1978), Campbell and Clark (1971), Clark and Evans (1954, 1955), Cooper (1961), Cowie (1967), Crisp (1961), Diggle (1976, 1979), Ellis et al. (1969), Gulmon and Mooney (1977), Hsu and Mason (1974), Kennedy and Crawley (1967), Lawrence (1969), Lesseps et al. (1975), Marquiss et al. (1978), Newton et al. (1977), Persson (1972), Pinder and Witherick (1972), Ripley (1977, 1979c), Roder (1975), Sands (1965), Sexton and Stalker (1961), Stimson (1974), Sukwong et al. (1971), Vincent et al. (1976b), Walloff and Blackith (1962), Ward and Sprontz (1976).

8.3   SECOND MOMENTS

We have seen (in Section 8.1) the connection between second moments of counts and the distribution of the distances between pairs of objects. Suppose first that we have a guard area, and let $\hat{F}$ be the empirical cumulative distribution function of distances from points in D to distinct points in D or in the guard area.   The size of the border must be sufficient to include all points not more than t_0 away from D.   The sample size is unclear, so suppose for the moment $\hat{F}$ has jumps of size 1.   Then, from (8.2),

$$E\left(\hat{F}(t)\right) = E(\text{number of pairs, one in } D, \text{ second within } t \text{ of first})$$

$$= A\lambda^2 K(t) \qquad \text{for } t \leqslant t_0 \tag{8.6}$$

Clearly, we should rescale by λ^{-2}, but λ is unknown.   An unbiased estimator of λ is

$$\hat{\lambda} = N/A$$

whereas for a Poisson process $N(N-1)/A^2$ is an unbiased estimator of λ^2. It is usual to consider

$$\hat{F}(t) = \left(\sum 1 \right) / N(N-1) \qquad (8.7)$$

the sum being over all *ordered* pairs of distinct points in D not more than t apart (Bartlett, 1964). Since we no longer allow a guard area, we expect a downward bias unless toroidal edge correction is applied, when (8.6) holds up to one-half the shorter side of the rectangle. The exact expectation for a binomial process (and for a Poisson process, because it does not depend on N) is known; see Kendall and Moran (1963, pp. 41–42) and Geciauskas (1977). However, it is possible to produce an unbiased estimator by weighting pairs of objects. We define (for $N \geqslant 2$)

$$\hat{K}(t) = A \left(\sum k(\mathbf{x}, \mathbf{y}) \right) / N^2 \qquad (8.8)$$

the sum being over the same pairs $(\mathbf{x}, \mathbf{y})$ as (8.7), but $1/k(\mathbf{x}, \mathbf{y})$ is defined as the proportion within D of the circumference of the ball centered on $\mathbf{x}$ with boundary passing through $\mathbf{y}$. Figure 8.5 explains the intuitive idea; since some objects at distance $d(\mathbf{x}, \mathbf{y})$ from $\mathbf{x}$ might be outside D and so be missed, we weight the observed object at $\mathbf{y}$ inversely by the probability that such an object would be observed. Of course we must consider both $(\mathbf{x}, \mathbf{y})$ and $(\mathbf{y}, \mathbf{x})$. We find (Ripley, 1976a)

$$E(N^2 \hat{K}(t)) = [E(N)]^2 K(t) = (\lambda A)^2 K(t) \qquad (8.9)$$

for distances t less than the circumradius of D. $\hat{K}(t)$ should not be highly correlated with N, so $\hat{K}(t)$ will be an approximately unbiased estimator of $K(t)$. Although the derivation of (8.9) via (8.2) assumed homogeneity and isotropy, we can find the mean of $\hat{K}(t)$ for a binomial process with N points. It is $\pi t^2(1 - 1/N)$ and is also the conditional mean $E(\hat{K}(t)|N(D) = N)$ for a Poisson process. The factor $(1 - 1/N)$ is related to the choice of divisor N^2 or $N(N-1)$; it seems immaterial in practice. From the interpretations of $K(t)$ given in Section 8.1 we would expect $\hat{K}(t)$ to be smaller than πt^2 (its approximate expectation for a Poisson process) if the data form an inhibited or regular pattern, and to be larger in the presence of clustering or heterogeneity. Furthermore, the sizes of the deviations from πt^2 and the distances at which they occur should give some further insights into the pattern. Besag, in the discussion of Ripley (1977), suggested the use of a square-root scale to linearize the plot of $\hat{K}(t)$ vs. t for a Poisson process. This also has the effect of stabilizing the variances for we shall see that $N^2 \hat{K}(t)/A$ has an approximately Poisson distribution. Note that this is a fortuitous coincidence, for whereas square roots always stabilize variances the linearizing transformation is the dth root in $\mathbb{R}^d$. Define

$$L(t) = \sqrt{(\hat{K}(t)/\pi)}$$

$$L_m = \sup_{t < t_0} |L(t) - t|$$

Then small values of L_m provide an intuitively reasonable test region for the null hypothesis of a Poisson process, for some maximum distance of interest t_0. Its distribution is approximated later in this section.

At this point the reader may wish to look ahead to Section 8.6 to see some examples of the use of $\hat{K}$ and L_m.

Ripley and Silverman (1978) introduced a family of statistics related to both those in the last section and in this. Let $d_{(1)} \leqslant d_{(2)} \leqslant \cdots$ be the distances $\{d_{ij}\}$ in increasing order. Then $d_{\min} = d_{(1)} = \sup\{t \mid \hat{K}(t) = 0\}$ is the smallest nearest-neighbor distance, and $L_m \geqslant d_{\min}$, with probable equality when the pattern is strongly regular. For a binomial or Poisson process

$$N(N-1)\pi d_{(i)}^2 / A$$

has a $\chi^2_{(2i)}$ distribution for practical purposes for $N \geqslant 15$ ($i=1$) or $N \geqslant 30$ ($i \leqslant 9$). In particular, $d_{\min}^2$ has an exponential distribution of mean $2A/\pi N(N-1)$. These statistics are vulnerable to recording errors and should only be used when the coordinates are known better than to within one-third of $d_{\min}$.

Both Jolivet (1978, 1980) and Liebetrau (1977, 1978, with Rothman, 1977) consider estimators of $E(N(C)^2)$ for a rectangle C of sides c_1 and c_2 and of area a. Liebetrau's approach is to consider C on a torus D and to

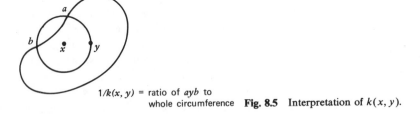

$1/k(x, y)$ = ratio of ayb to whole circumference   **Fig. 8.5** Interpretation of $k(x, y)$.

average $[N(C+\mathbf{x})-a/A]^2$ over all translations $C+\mathbf{x}$ of C. The average is

$$V(C)=a\sum \phi(\mathbf{x},\mathbf{y})/A+Na/A-N^2a^2/A^2 \qquad (8.10)$$

where the sum is over all ordered pairs $(\mathbf{x},\mathbf{y})$ of distinct points and

$$\phi((x_1,x_2),(y_1,y_2))=(c_1-|x_1-y_1|)_+(c_2-|x_2-y_2|)_+/c_1c_2$$

$$(t)_+=\max(t,0)$$

For a Poisson process $V^*=V(C)/(1-a/A)$ is an unbiased estimator of $\mathrm{var}(N(C))=\lambda a$. A test of a Poisson process can thus be based on the ratio

$$L^*=(V^*/\hat{\lambda}a)-1 \qquad (8.11)$$

an analogue of David and Moore's index of Section 6.1 computed from all possible quadrat positions rather than from a finite sample.

Jolivet (1978) defined $\mathcal{C}(C)=E(N(C)^2)/E(N(C))$ as a "criterion for aggregation" and estimated this by

$$J=\sum \frac{\phi(\mathbf{x},\mathbf{y})}{\lambda A}+\frac{N}{\lambda A} \qquad (8.12)$$

where the sum is over ordered pairs of distinct points $(\mathbf{x},\mathbf{y})$ with $\mathbf{x}\in D$, but where $\mathbf{y}$ is allowed in a guard area. The second term arises because Jolivet counted the pairs $(\mathbf{x},\mathbf{x})$. J is unbiased if λ is assumed known.

Distribution Theory

Note that $\hat{F}$, $\hat{K}$, V, and J can all be written in the form

$$T=a\sum \psi(\mathbf{x},\mathbf{y})+b$$

where $\mathbf{x}\in D$ and either $\mathbf{y}\in D$ or $\mathbf{y}\in D^*\supset D$ if we allow a border. Thus we can treat their asymptotic behavior in a unified way. Many asymptotic possibilities are open, for we can vary N (or λ), D, and C or the scale of the distances separately or together. Perhaps the most natural case, considered by Liebetrau (1978) and Jolivet (1978, 1980) is to allow D to expand and to include more and more of a realization of a point process throughout the plane. Many authors have considered the asymptotic behavior of $\hat{F}(t)$ for a binomial or Poisson process: Hafner (1972), Kester

(1975), Silverman (1976, 1978), Saunders and Funk (1977), Silverman and Brown (1978), and Brown and Silverman (1979).

Consider a binomial process with N objects and no border, in a square of side d. Let $\chi(\mathbf{x}, \mathbf{y}) = \frac{1}{2}\{\psi(\mathbf{x}, \mathbf{y}) + \psi(\mathbf{y}, \mathbf{x})\}$ so

$$T = 2a \sum_{i<j} \chi(X_i, X_j) + b \tag{8.13}$$

where (X_i) are independent uniformly distributed random variables on D. Equation (8.13) exhibits T as a U-statistic (Lehmann, 1975, Appendix 5). However, we will allow χ and the distribution of X_1 to change with N, making the standard asymptotic theory for U-statistics inappropriate. We find

$$E(T) = aN(N-1)E(\chi(X_1, X_2)) + b$$

$$\mathrm{var}(T) = 4a^2N(N-1)\Big[(N-2)\,\mathrm{var}\{E(\chi(X_1, X_2)|X_1)\}$$

$$+ \tfrac{1}{2}\,\mathrm{var}(\chi(X_1, X_2))\Big] \tag{8.14}$$

The first term in (8.14) is due entirely to edge effects, being zero when toroidal edge correction is used.

Consider first $\hat{F}$ and $\hat{K}$. For small t/d we have (Ripley, 1979b)

$$E(\chi(X_1, X_2)) \approx \pi(t/d)^2$$

$$\mathrm{var}(\chi(X_1, X_2)) \approx E(\chi(X_1, X_2))$$

$$\mathrm{var}\big[E(\chi(X_1, X_2)|X_1)\big] \approx 4\kappa(t/d)^5 \qquad \kappa = 0.672 \text{ for } \hat{F}$$

$$0.061 \text{ for } \hat{K}$$

$U = \sum_{i<j} \chi(X_i, X_j)$ is a count for $\hat{F}$, and an approximation to a count for $\hat{K}$, with mean and variance

$$E(U) \approx (Nt/d)^2 \pi/2$$

$$\mathrm{var}(U) \approx (Nt/d)^2 \pi/2 + 4\kappa(Nt/d)^5/N^2 \tag{8.15}$$

Clearly, Nt/d is a crucial parameter. If it is small, we have a count of rare events and will consider a Poisson approximation provided the second variance term is negligible compared with the first. Thus, for 25 points, a Poisson approximation is suitable for $t/d \leqslant 0.25$ for $\hat{K}$, 0.13 for $\hat{F}$, and for

$N \geqslant 50$, for $t \leqslant 6d/N$ for $\hat{F}$ or $\hat{K}$. Beyond these limits we must use a Normal approximation. Note that in the range of t for which the first term of (8.15) dominates the second var$(\hat{F})$ and var$(\hat{K})$ are proportional to N^{-2}. The terms become equal at $t=0.84d/N^{1/3}$ for $\hat{F}$, $t=1.86\,d/N^{1/3}$ for $\hat{K}$. In particular, for $t \geqslant 0.6\,d/N^{1/3}$, $\hat{K}(t)$ has an appreciably smaller variance than $\hat{F}$. This value is two to three times the mean nearest-neighbor distance for usual values of N. These Poisson and Normal approximations are supported by limit theorems, which do, however, give a warning; with toroidal edge correction, t, D fixed and $N \to \infty$ the limiting distribution of $N(\hat{F}(t)-F(t))$ is non-Normal (Silverman, 1978; Gregory, 1977).

Not only can we regard $N^2\hat{K}(t)/2A$ as approximately Poisson for $t \leqslant 0.9d/N^{1/3}$; we can approximate $N^2\hat{K}/2A$ by a Poisson process of intensity $N(N-1)2\pi t/A$ in this range. This result gives the distribution of $d_{(i)}$ and can also be used to find percentage points for L_m by simulating this Poisson process.

We can apply a similar analysis to

$$I=2\sum_{i<j} \phi(X_i, X_j)$$

except that since this is not a count, a Poisson limit is inappropriate. We have, for a binomial process with N points and $e=$ area $(C)/$area $(D)=a/A \ll 1$,

$$E(I)=N(N-1)e$$

$$\mathrm{var}(I)\approx N^2(8e/9-2e^2)+N^3(3.2e^{2.5})$$

for var$[E\{\phi(X_1, X_2)|X_1\}]\approx 0.8e^{2.5}$.

We can safely ignore the edge effect term if $e \leqslant N^{-2/3}/2$, and of course if toroidal or guard area edge correction is used. In that case

$$E(V^*)=Ne \qquad \mathrm{var}(V^*)\approx 8e^3N^2/9$$

$$E(L^*)=0 \qquad \mathrm{var}(L^*)\approx 8e/9$$

$$E(J)=[N+N(N-1)e]/\lambda A$$

$$\mathrm{var}(J)\approx 8N^2e/9(\lambda A)^2$$

Both Liebetrau and Jolivet considered Poisson processes, so we take expectations over N. Note that the mean and variance formulas for L^*

are independent of N. For J we have

$$E(J) = 1 + \lambda a$$

$$\text{var}(J) = (4 + 8/9)e + 4\lambda ae + 1/\lambda A$$

so the variance for a Poisson process is at least five times that for a binomial process with the same intensity! Clearly, the conditional variance should be used.

8.4 MODELS

Chapter 6 introduced two informal models for clustering and for heterogeneity and pointed out the tenuous nature of the distinction between them. We can now formalize these processes and show that certain distributions of point processes can arise from either mechanism.

Cluster Processes

A Poisson cluster process is defined by taking a Poisson process of intensity α of parent points and centering on each parent an independent daughter process of objects. The observed process may be either parents plus daughters or just all daughter objects. We will assume the latter. Another way to view this mechanism is to have an infinite set of independent identically distributed processes, to use a Poisson process to select a translation for each, and then to add up the objects in all the translated processes. This suggests a modification in which we choose a Poisson process of rigid motions, thereby giving each daughter process an independent uniformly distributed rotation. We will assume that each daughter process contains a finite number of objects. The cluster process will always be homogeneous, but will only be isotropic if the daughter process is isotropic or if the daughters are given an additional rotation. The most useful subclass of Poisson cluster processes is Neyman–Scott processes, for which each daughter object is independently distributed around the parent. Then if n is the (random) number of objects in the daughter process,

$$K(t) = \pi t^2 + \alpha E(n(n-1))f(t)/\lambda^2 \tag{8.16}$$

$$p(t) = 1 - \exp\left\{ -\alpha \int \left[1 - E\{g(\mathbf{x}, t)^n\} \right] d\mathbf{x} \right\} \tag{8.17}$$

where all expectations are over n, f is the cumulative distribution function of the distance between two daughters with the same parent, and $g(\mathbf{x}, t)$ is the probability that a daughter point does not fall within distance t of $\mathbf{x}$.

Equation (8.16) can be derived from interpretation (1) of $K(t)$. The pairs of points can come either from different clusters giving the first term by the independence of clusters, or from the same cluster. If there are n objects in that cluster, the expected number of pairs not more than t apart is $n(n-1)f(t)$, from which is derived the second term. Note that $K(t) - \pi t^2$ is an increasing function and that we can infer an estimate of $f(t)$, and hence the cluster size, from $\hat{K}(t)$. Formula (8.17) is a special case of (9.10). Consider the N parent points within a bounded set D. For large D of area A

$$1 - p(t) \approx P(\text{no daughter with parent in } D \text{ is within } t \text{ of the origin})$$

$$= \sum e^{-\alpha A}(\alpha A)^N \{P(\text{no daughter within } t | \text{parent in } D)\}^N / N!$$

$$= \exp{-\alpha A}\{1 - P(\text{no daughter within } t | \text{parent in } D)\}$$

$$= \exp{-\alpha A}\left\{1 - \int_D E(g(\mathbf{x}, t)^n) \, d\mathbf{x}/A\right\}$$

using the independence of the daughters and the uniform distribution of the parent in D. Letting D increase gives (8.17).

Of course the parent process need not be Poisson; it could itself be a cluster process, giving rise to processes with a (finite) hierarchy of clusters. Another possibility is to take a regular process of parents to avoid the overlap of clusters.

Doubly Stochastic Poisson Processes

We can define a homogeneous process by taking a heterogeneous Poisson process with mean measure Λ, where Λ is a realization of a stochastic process, stationary under translations. If Λ is isotropic, then so is the resulting point process, called Cox or doubly stochastic Poisson. As an example, suppose we take the Neyman-Scott cluster process with a Poisson number mean β of daughters uniformly distributed within a disk of radius r centered on the parent. Then if we condition on the parent Poisson process, we have a Poisson process of intensity $\lambda(\mathbf{x})$, β times the number of discs which cover $\mathbf{x}$. Thus this is a Cox process, one of several defined by Matérn (1971). Clearly, this duality holds whenever the number of daughters in a Neyman–Scott cluster process has a Poisson distribution.

It seems unknown just how generally Poisson cluster processes are also Cox processes. (Bartlett, 1963, gives a counterexample in one dimension.)

Regular Processes

We introduced in Section 2.4 a general way to build new point processes from the Poisson process. The most useful are those based on pair potentials, with

$$\phi(\mathbf{x}) = ab^{\#(\mathbf{x})} \prod_{i<j} h\big(d(x_i, x_j)\big) \qquad (8.18)$$

and the probability of any event E is

$$P_\phi(E) = \int_E \phi(\mathbf{x}) \, dP_0(\mathbf{x})$$

where P_0 is the distribution of a Poisson process of intensity λ. As noted in Section 2.5, such processes can only be defined directly on a bounded set D. We can and usually will sidestep problems at the edges by making D a torus, opened up to be a rectangle. The constant a is chosen to make $P_\phi(\Omega) = 1$, so a is almost always an unknown function of b, λ, and the parameters defining h.

Kelly and Ripley (1976), following Strauss (1975a), introduced Strauss processes with $h(d) = c, d < r, h(d) = 1, d > r$ so that

$$\phi(\mathbf{x}) = ab^{\#(\mathbf{x})} c^{t(\mathbf{x})} \qquad (8.19)$$

where $t(\mathbf{x})$ is the number of pairs of points in $\mathbf{x}$ less than r apart. The necessary and sufficient condition for a to be able to be chosen finite and nonzero is $0 \leqslant c \leqslant 1$. This family is characterized by the property that the probability density of an object being found at a point x, conditional on all the other objects, depends only on the number of objects within r of x. An important special case is $c = 0$, which corresponds to keeping only those realizations of a Poisson process with no object pairs less than r apart, and so is a model for the centers of nonoverlapping disks of radius $(r/2)$.

Strauss processes have a subtly different sequential packing version. The first object is placed uniformly in D. Subsequent objects are generated uniformly in D, and accepted with probability c^s, where s is the number of existing objects closer than r to the possible new object. This process is repeated until a fixed number of objects are tried, or a given number are accepted. For $c = 0$ the latter is the *SSI* process of Smalley

(1966) and Diggle et al. (1976). Other, related, processes for disk centers are described by Matérn (1960), Paloheimo (1971), Bartlett (1974), and Ripley (1977).

Ord (in the discussion of Ripley, 1977) proposed a modification in which the areas of the Dirichlet cells were used. This seemed appropriate for towns and possibly trees when a reserved food area is needed. This model would have a density of the form

$$\phi(\mathbf{x}) = ab^{\#(\mathbf{x})} \prod_i g(\text{area of Dirichlet cell of the } i\text{th point}) \qquad (8.20)$$

Almost no analytical work is possible with the models defined by (8.18) to (8.20) and we have to resort to simulation. In principle, a realization of the process can be obtained by sampling realizations of a Poisson process and accepting each realization independently with probability $\phi(\mathbf{x})/M$, where M is an upper bound for $\phi(\mathbf{x})$, if such exists. In practice, almost every realization is rejected, making this process far too slow. An alternative simulation technique has been proposed by Ripley (1977, 1979a). To generate a sample in D with N objects take any pattern with N points and $\phi(\mathbf{x}) > 0$ and operate the following process repeatedly. Choose one object at random, delete it, keeping $(x_2, \ldots, x_N)$, say, and choose a replacement object x_1 from the conditional density

$$A\phi(x_1, \ldots, x_n)/\phi(x_2, \ldots, x_n)$$
$$= Ab \prod_2^n h(d(x_i, x_1)) \qquad \text{for (8.18)}$$

Again the normalizing constant A is chosen to make this a probability density. We probably still do not know A, but the rejection procedure suggested as impracticable for all N points may be possible for x_1 conditional on $(x_2, \ldots, x_N)$. This process ultimately yields (dependent) samples from the distribution P_ϕ. With a sensible choice of initial pattern practically independent samples can be taken every $2N$ steps. A FORTRAN algorithm for the Strauss process and proofs are given in (Ripley, 1979a).

Estimation

Estimating the parameters for these processes is a difficult problem. Little direct progress has been made towards its solution. For cluster processes maximum likelihood involves looking at all possible allocations of objects to clusters. For pair-potential processes the likelihood contains an unknown normalizing constant which is a function of the parameters, so progress by classical means is immediately blocked. Besag (1978) estimated the parameter c of a Strauss process by taking a grid of counts and fitting an auto-Poisson process to this by the approximate method of pseudo-likelihood. The resulting estimates for the Spanish towns data of Ripley (1977) fit less well than do those obtained in that paper by trial and error. For this model with $c = 0$ Ripley and Silverman (1978) showed d_{min} is a sufficient statistic for r and that a UMP test for $r = 0$ (Poisson) against $r > 0$ is based on d_{min}. Further, the maximum likelihood estimate of r is d_{min} and the asymptotic distribution theory for d_{min} can be used to give confidence limits for r, for $N(N-1)(d_{min}^2 - r^2)$ has approximately an exponential distribution of mean $2A/\pi$.

Diggle (1978, 1979, 1980) has formalized the trial and error procedure. A goodness-of-fit test will be based on some "distance" between a statistic T for the data and its expectation T^θ for the model with parameter θ. Diggle's idea amounts to choosing θ to minimize this "distance" and then choosing some other statistic as a test of goodness of fit. If T^θ is not known it can be estimated by simulations of the model. Such a procedure is frequently used informally. A better check on the fit of the chosen model can be made by looking at more data. Frequently extra data are unavailable. One way out is to divide the original data into two or more subsamples and to estimate parameters from one sample, testing the fit against the rest. In practice commonsense will prevail. Often the problem is to find a good enough fit with parameters estimated from the data!

8.5 COMPARATIVE STUDIES

Both Diggle (1979) and Ripley (1979b) have published power studies of selections of the tests described earlier for the null hypothesis of a binomial process against cluster, Strauss, and SSI alternatives with the same number of objects. Diggle performed 4% Monte Carlo tests with $N = 100$ objects based on the same 99 binomial simulations. He considered the Clark-Evans test CE, d_x, d_y, d_2, L_m (but with $t_0 = 0.25d$, unreasonably large for $N = 100$) and the variance/mean ratio q for a 5×5 grid of squares. Against SSI alternatives L_m was clearly dominant. The cluster alternative was a Poisson cluster process with one plus a Poisson number in each cluster, with mean ranging from two to four daughters per cluster. Here all of CE, L_m, d_x, and q did well. Perry and Mead (1979) give some additional power studies for q.

Ripley (1979b) considered CE, d^*, S, G, L_m and I (equivalent to J and L^* for a fixed sample size). Against the regular alternative of a Strauss process L_m (with $t_0/d = 0.25$ for $N = 25$, 0.125 for $N = 100$) was clearly superior, with CE next and G effective against packed disk alternatives, that is, $c = 0$. For cluster processes of the type described under "doubly stochastic Poisson processes," CE, L_m, and I were equally effective, with I surprisingly effective at large numbers in large clusters that would normally be classified as heterogeneity. For I the square C was taken as $t_0 \times t_0$; thus it corresponds to an analogue of q with all quadrat positions considered (and quadrats of about the same size as those used in Diggle's study).

We have seen how toroidal edge correction can simplify the distribution theory for methods based on nearest neighbors and on second moments. The Ripley (1979b) study was conducted both with specialized edge corrections (Donnelly, $\hat{K}$) and with toroidal edge correction. There was no significant difference in power against cluster alternatives; against regular alternatives the toroidal correction was usually less effective (giving lower power); this was most marked for L_m.

These seem the only studies on which one can base recommendations. Tests based on CE, L_m, and I have the clear advantage that their distribution under the binomial process is known to a good approximation, at least for nearly square regions D. These are also the tests that did well in the power comparison. If hand calculation is necessary, only CE would be used. (However, it is surprisingly difficult to find all the nearest-neighbor distances correctly in a pattern of any size.) With a computer all three methods involve searching over all pairs of points and so take approximately the same time, proportional to N^2. More efficient methods are available for large N. For nearest-neighbor distances we could use the Dirichlet cell contiguity information, since the nearest neighbor must be the defining object of a contiguous cell. Ultimately, this needs a time of order $N \log N$. For L_m and I we will be interested in short distances, and by sorting the data objects we can screen out a large number of pairs. However, at least in my implementations, none of these complications is as fast as the naive methods for $N \leqslant 100$.

8.6 EXAMPLES

Other examples of analyses of point patterns by similar methods may be found in Ripley (1977, 1979c). The former are considered further by Besag (1978) and Diggle (1978, 1979, 1980).

New Zealand Trees

Figure 8.6 illustrates the analysis of a plot of trees reported by Mark and Esler (1970). These workers recorded the diameter of each tree and

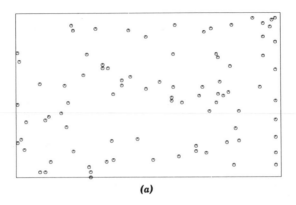

(a)

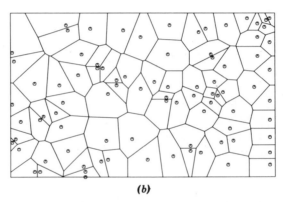

(b)

Fig. 8.6 New Zealand trees–86 points in a 140×85 feet rectangle. (*b*) Dirichlet cells. (*c*) *L*. (*d*) $\hat{p}$ for the data and the envelopes of $\hat{p}$ for 50 binomial simulations.

related this to the "area potentially available" of Brown (1965), the area of the Dirichlet cell defined by the tree. We will be concerned solely with the pattern of the trees as small objects. There are 86 trees in a rectangle about 140×85 feet. There is an obvious boundary line on the right-hand edge of the plot in 8.6*a*. These trees were discarded from the analysis. Figure 8.6*b* shows the Dirichlet cells and illustrates the number of close pairs of trees. The statistics *L* and $\hat{p}$ are shown in Figure 8.6*c*, *d*. The band on the plot of *L* is the 95% confidence band based on L_m; 95% of realizations of *L* calculated from a binomial process should lie within this band. *L* for the data does! The minimum distance between a pair of trees is about 2 feet, near the upper 5% point for a binomial process, but

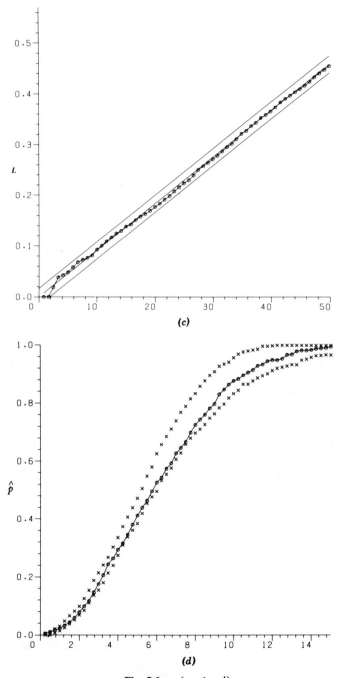

(c)

(d)

Fig. 8.6 (*continued*)

N.B. The y-axis scale of all L plots shows L/√A.

171

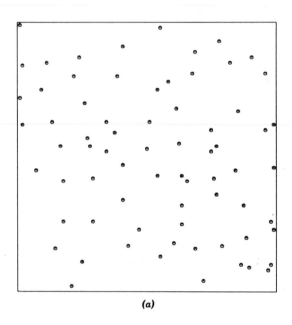

(a)

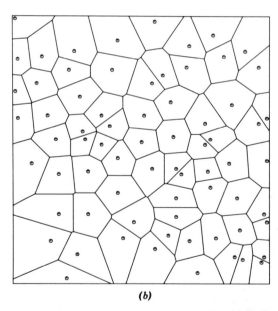

(b)

Fig. 8.7 Swedish pines. 71 trees in a 10-meter square. (*b*) Dirichlet cells. (*c*) *L*. (*d*) $\hat{p}$ for data and 100 binomial simulations. (*e*) and (*f*) *L* and $\hat{p}$ for data and 50 simulations of a Strauss process with $c=0.2$, $r=70$ cm.

172

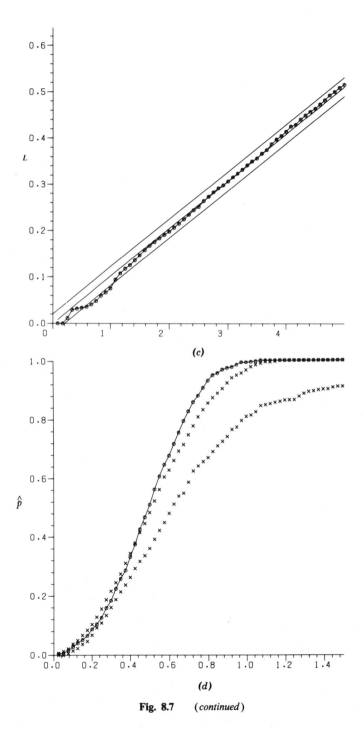

(c)

(d)

Fig. 8.7 (*continued*)

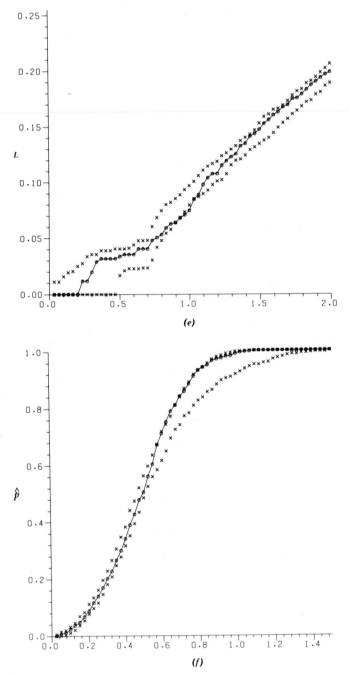

(e)

(f)

Fig. 8.7 (*continued*)

unreliable since the data were digitized to the nearest foot. The $\hat{p}$ plot shows the upper and lower envelopes of $\hat{p}$ for 50 binomial simulations. Since a Monte Carlo test shows that at each distance t the probability of the data giving an extreme value of $\hat{p}(t)$ is $2/51$, and this never occurred, clearly this data set is consistent with a binomial process, and hence with a Poisson process.

Swedish Pine Trees

Not all trees have a random pattern. Our next example, Figure 8.7a, shows 71 pine saplings in a 10×10-meter square from Strand (1972). The Dirichlet cells in Figure 8.7b look much more regular than do those of the previous example, suggesting that the pattern is on the regular side of random. The plot of L in Figure 8.7c rejects the Poisson null hypothesis (in fact, L_m does so at 1%), and d_{min} is 22 cm, at the upper 2% point, again suspect because of insufficient accuracy in the coordinates. The $\hat{p}$ analysis of Figure 8.7d again suggests that the trees are regularly spaced, there being too few empty circles with a 70-cm radius.

Having rejected a Poisson process, we look for an alternative. The plot of L suggests we might choose a Strauss process with an inhibition distance $r = 70$ cm, about the point of maximum deviation of L from the central straight line. The value of c was chosen by trial and error. Figure 8.7e, f compares L and $\hat{p}$ for the data and 50 simulations of the Strauss process with $r = 70$ cm, $c = 0.2$. Although the $\hat{p}$ plot suggests that the data may be more regular than these simulations, the fit is much more adequate than a Poisson process. The value of 70 cm makes physical sense for young trees. An alternative model would be to use Ord's suggestions and to discriminate against small Dirichlet cell areas. The parameters of the Strauss process provide a convenient summary of the data, which may be compared with those from similar plots elsewhere.

Magnetite Crystals

Davis (1973, pp. 308–311) gives an example of 47 magnetite crystals in a vertical 80×100-cm slab of anorthosite rock. He detected regularity spuriously, using the uncorrected version of Clark-Evans's test. (Donnelly's corrections reduced the value of CE from 2.59 to 1.48.) Figure 8.8a shows an apparent increase in intensity from top to bottom. Whether this is significant will depend on the type of pattern, for we would expect large intensity variations in cluster processes, but almost no variation in a very regular process. Initially we will suppose that the underlying process

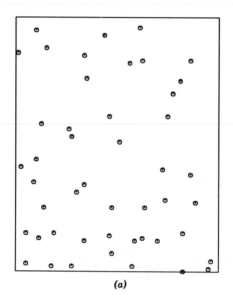

(a)

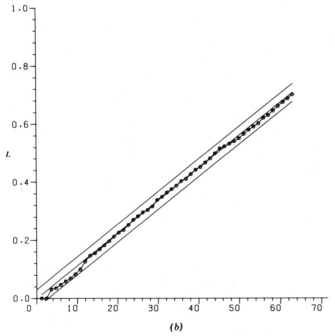

(b)

Fig. 8.8 Magnetite crystals. 47 in a 80×100 cm slab of rock. (b) L. (c) and (d), L and $\hat{p}$ for data and envelope of 100 binomial simulations.

176

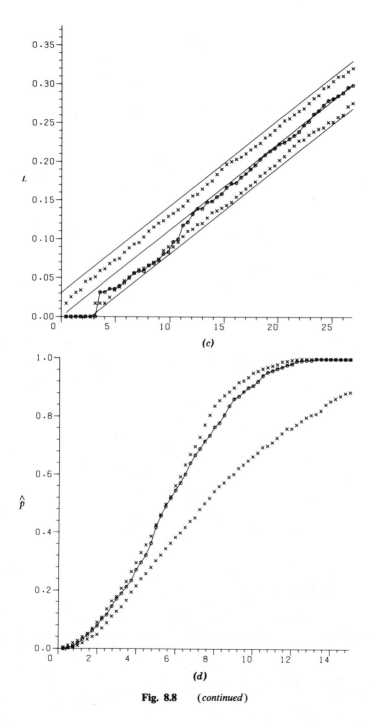

Fig. 8.8 (*continued*)

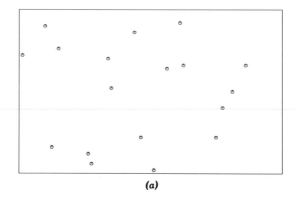

(a)

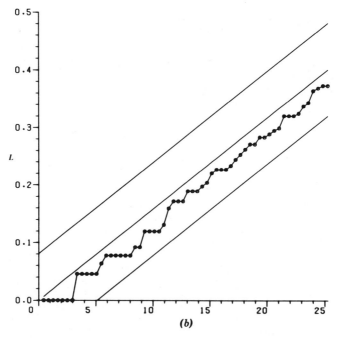

(b)

Fig. 8.9 Top (a, c) and bottom (b, d) halves of magnetite crystal data.

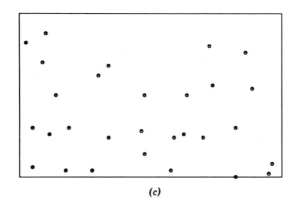

(c)

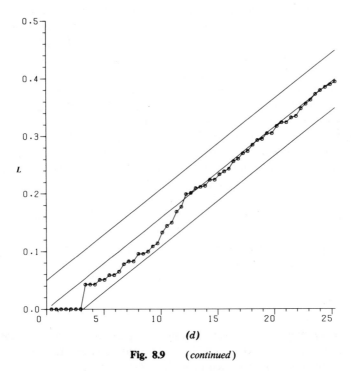

(d)

Fig. 8.9 (*continued*)

179

is homogeneous.   Figure 8.8b, c shows plots of L.   From the enlargement in Figure 8.8c we see that L_m rejects the Poisson null hypothesis at 5% solely because d_{min} is too large.   (The envelope of simulations shown there shows that the confidence band is of the correct shape.)   In fact, d_{min} is 3.2 cm, about the upper 1% point of its null distribution.   The points were recorded to the nearest centimeter, so this should be reliable.   The plot of $\hat{p}$ in Figure 8.8d does not reject the Poisson null hypothesis, but is consistent with slight inhibition.

Figure 8.9 shows the upper and lower halves of the slab separately. These contain 18 and 29 crystals.   From the previous analysis we will be justified in regarding these as independent Poisson variates.   A chi-square test for the difference in their means has a significance level of about 10%. Many other trends in intensity could have been spotted, so we attribute the intensity variation to chance.   The plots of L for the two halves are reassuringly similar.

Apparently some inhibition between crystals would be expected since they compete for heat energy during the growth process.   Our next step is to fit a hard disk process to model this.   We chose the Strauss model with $c = 0$.   It remains to choose the disk diameter r.   Section 8.4 gave the maximum likelihood estimator as d_{min}; we chose r as the *median* of the exponential distribution for $d_{min}^2 - r^2$, giving $\hat{r} = 2.9$.   Figure 8.10 compares L and $\hat{p}$ for the data and 100 simulations of this process.

There are more sophisticated techniques available for detecting trends in a Poisson process.   If we are looking for a trend in the y direction, the y coordinates of the points form a Poisson point process on the line to which we can apply the tests described in Cox and Lewis (1966, Chapter 3). Also, we saw in Chapter 6 that we can apply log-linear models to quadrat counts.   Using 10 horizontal rectangles and a trend of the form $e^{+\alpha y}$, α is estimated as 1.06 per meter, and the likelihood ratio test for $\alpha = 0$ has a value 4.27 of the "deviance," to be referred to a χ_1^2 distribution.   This suggests that the trend is somewhat more significant than does the binary division.   It appears that the bottom 20 cm has a rather higher intensity than does the rest of the rock.

Birds' Nests

Tubbs (1974), Newton et al. (1977), and Marquiss et al. (1978) analyze the spatial distribution of nest sites of buzzards, sparrowhawks, and ravens, all large birds of prey, the first two of which nest in tracts of woodland.   All three studies worked with nearest-neighbor distances and had problems with the disconnected nature of the study area.   Here we look at two

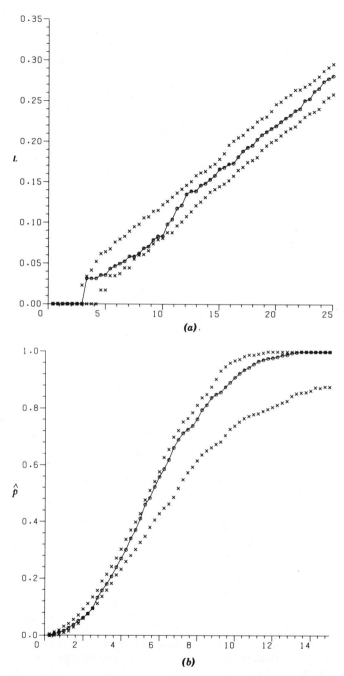

Fig. 8.10 L and $\hat{p}$ for crystal data and 100 simulations of hard disk model with disk diameter 2.9 cm.

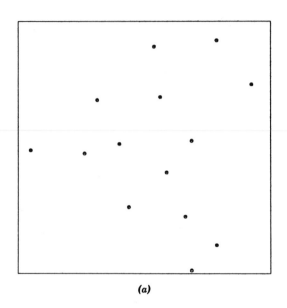

(a)

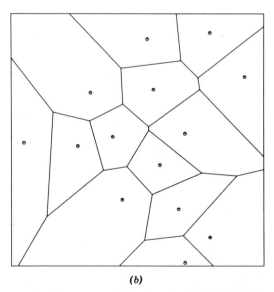

(b)

Fig. 8.11 Eagles' nest sites in a square of side 80 units. (*b*) Dirichlet cells. (*c*) *L*. (*d*) $\hat{p}$ plus the envelopes of 100 binomial simulations. (*e*) and (*f*) *L* and $\hat{p}$ for data and 100 simulations of hard disk model diameter 10.5 units.

182

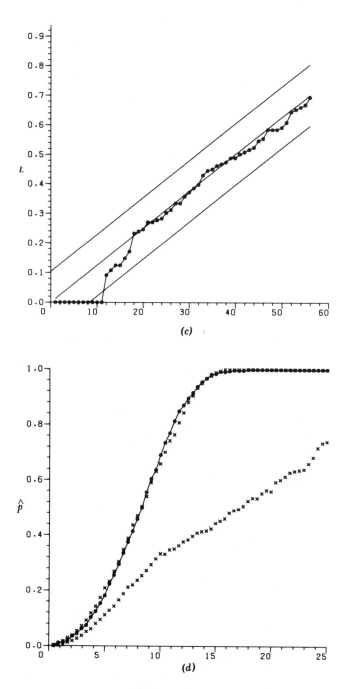

Fig. 8.11   (*continued*)

183

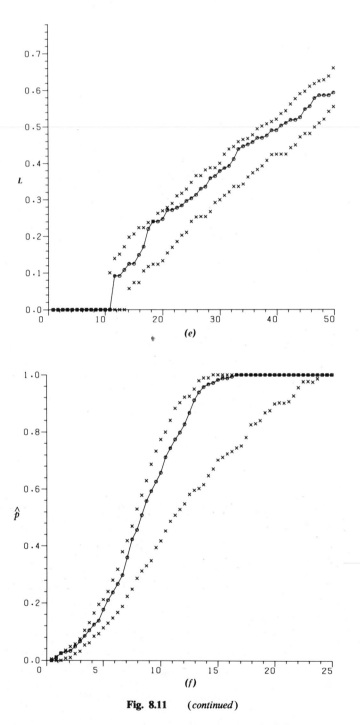

(e)

(f)

Fig. 8.11   (*continued*)

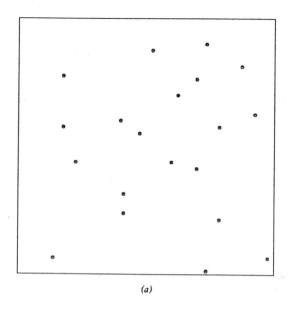

(a)

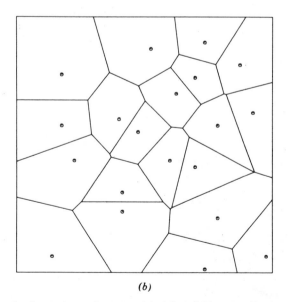

(b)

Fig. 8.12 Peregrines' nests in a unit square. (a)–(f) as 8.11 except diameter 0.075. (g) and (h) L and $\hat{p}$ for data and 50 simulations of the model with pair function shown at (i).

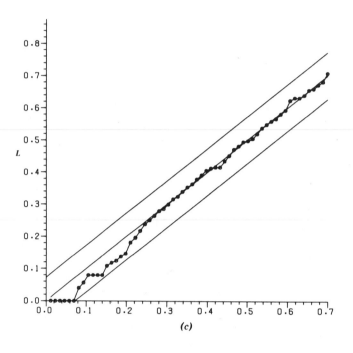

(c)

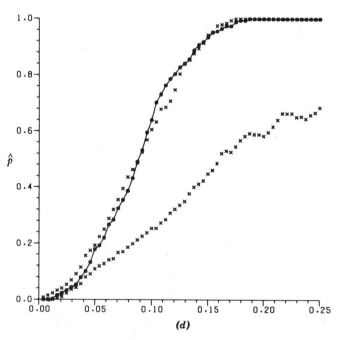

(d)

Fig. 8.12   (*continued*)

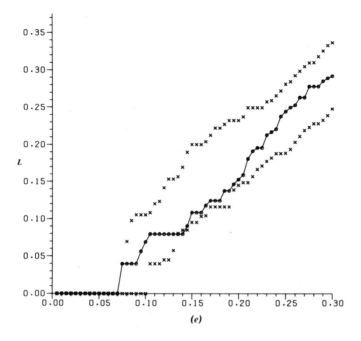

(e)

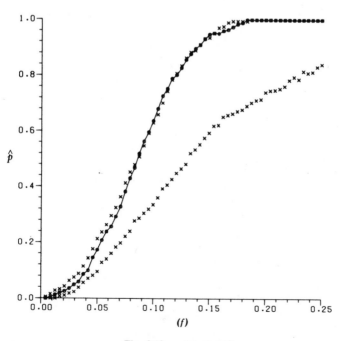

(f)

Fig. 8.12 (*continued*)

187

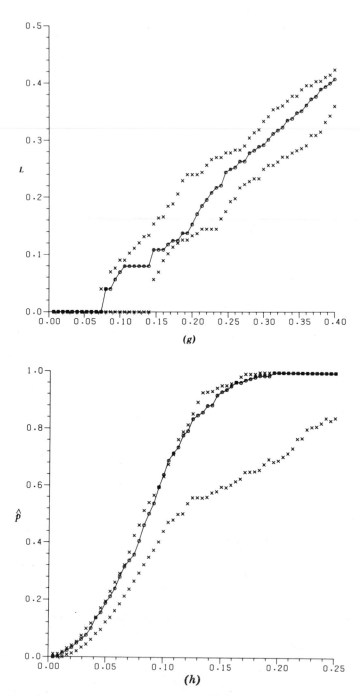

(g)

(h)

Fig. 8.12 (*continued*)

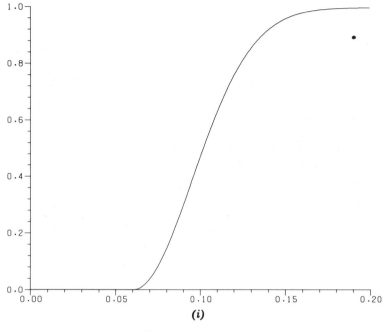

Fig. 8.12   (*continued*)

moorland nesting species—golden eagles and peregrines.   Figures 8.11*a*
and 8.12*a* illustrate the positions of the nests of 14 and 20 pairs of birds.
(Where a pair used a group of nests, the centroid has been chosen.)   The
scales and orientations are arbitrary.   It is a matter of dispute as to how
homogeneous the study areas are, and in particular whether we measure
the birds' preferences or the pattern of available nest sites.

For the eagles even with only 14 nest sites, the Dirichlet cells, L, and $\hat{p}$
all clearly indicate a regular pattern.   The minimum internest distance is
beyond the 1% point for a Poisson process.   Note that heterogeneity
would tend to reduce the true area and so reduce d_{min}.   The number of
nests is too few to use the asymptotic theory to choose the disk diameter r
as we did for the crystals.   In fact, r was chosen by trial and error as 10.5,
to give d_{min} for the data about the median for the simulations.   Figure
8.11*e,f* shows L and $\hat{p}$ for the simulations.   This model inhibiting nest
pairs closer than 10.5 units seems a good fit.   If we think of a circular
territory around the nest of this diameter, these territories cover only 19%

of the available area.   The existence of inhibition between nests is biologically somewhat surprising, as eagles hunt many miles from their nests and are not known to fight for territory.   Perhaps a better explanation is that we detect the pattern of outcropping rocks, some of which are occupied as nest sites.

The peregrines also exhibit obvious regularity, although L_m is only just beyond its 5% point.   Figure 8.12d, e illustrates an experiment with a hard disk model with $r = d_{min}$.   This is not too satisfactory, the shape of L indicating further discouragement against establishing nests just beyond that distance.   Figure 8.12i illustrates a more general pair potential function h of (8.18) designed to take this into account.   The corresponding simulations in Figure 8.12g, h show a slightly better fit.   The behavior of $\hat{p}$ in both these examples is puzzling, since it indicates a more regular pattern than that produced by the simulations that match the internest distance distribution.   It is also surprising that it is difficult to fit such small patterns.   Possibly there is some unexplained heterogeneity that affects $\hat{p}$.

CHAPTER 9

Image Analysis and Stereology

Figure 1.3 illustrates a complex type of pattern which we will model not by a random surface or set of points, but by a random *set*. Most of the applications are to three-dimensional bodies with two phases:

Bone and tissue
Blood cells and plasma
Pores in sedimentary rocks
Dislocations in metals
Glass fibers in resin

There are a few planar examples, often referred to as mosaics, such as the division of a region into different plant communities.

Practical measurements can be made either on planar slices or linear probes through a three-dimensional specimen. Even planar mosaics have until recently only been analyzed by a line intercept. Stereology is the theory of recovering three-dimensional information from one- or two-dimensional samples. Chayes (1972) provides an entertaining account of the early development of the subject. There may be problems even in quantifying a planar section. Image analyzers are essentially machines to scan a section and to record information at each point of a very fine grid (typically 1024×1024, $1024 = 2^{10}$). The information is usually a gray level or whether or not the density of the image exceeds a threshold. Clearly, even with a binary measurement a vast amount of information ($128K$ bytes) is available that can be handled by a dedicated minicomputer or stored on magnetic tape and analyzed later by a large computer. With a dedicated machine there is another possibility; some analyzers are addressable so the computer can ask for measurements at specific grid locations and so avoid the storage requirement. With the rapid decline in the cost of small computers and especially of memory this high-technology approach may become more widely available.

Random set theory was developed by Matheron (1967, 1975) to formalize the measurements which can be made by image analyzers and to provide some models for random sets. A more abstract version of the theory was developed independently by Kendall (1974). To date very little statistics (in the sense of the examination of and fitting of models to data) has been attempted. Section 9.1 sketches the probability theory, using a little topology.

The twin subjects of geometrical probability and integral geometry are much involved in the sampling theory of linear and planar sections of three dimensional bodies. Sections 9.2–9.4 consider aspects of this problem. Again, statistical aspects have been inadequately explored, for geometrical probability is well suited to proving unbiasedness under randomization but not to finding even standard errors.

9.1   RANDOM SET THEORY

Let X denote the basic space, usually $\mathbb{R}^2$ or $\mathbb{R}^3$, and consider a set A within X. For definiteness we will adopt the terminology of sedimentary rocks —A will be the conglomerate of grains and $X \backslash A$ the pore space. The most obvious way to specify a stochastic process Z generating A is by $Z(\mathbf{x}) = I_A(\mathbf{x})$, defined to be one if $\mathbf{x} \in A$, zero otherwise. (I_A is the indicator function of A.) Conversely, if Z is a function on X taking values zero or one, we can define the corresponding set A by

$$A = \{\mathbf{x} \,|\, Z(\mathbf{x}) = 1\}$$

The distribution of the stochastic process Z then specifies the probabilities of events of the form

$$\{Z(\mathbf{x}) = 1 \quad \forall \mathbf{x} \in K, Z(\mathbf{x}) = 0 \quad \forall \mathbf{x} \in K'\}$$

$$= \{A \supset K, \quad A \cap K' = \varnothing\} \tag{9.1}$$

for *finite* sets K and K'. This was the theory proposed by Matheron (1967), but it is insufficiently rich. To see why we have to analyze in more detail the functions of an image analyzer. Recall that the output may be considered as I_A evaluated at a very fine grid of points.

Basic operations:

Minkowski addition	$A \oplus B = \{x + y \mid x \in A, y \in B\}$
Reflection	$\check{B} = \{-x \mid x \in B\}$
Subtraction	$A \ominus B = (A^c \oplus B)^c = \{x \mid (x - B) \subset A\}$
Dilation of A by B	$A^B = A \oplus \check{B} = \{x \mid (B + x) \cap A \neq \varnothing\}$
Erosion of A by B	$A_B = A \ominus \check{B} = \{x \mid (B + x) \subset A\}$
Opening of A by B	$A \omega B = (A \ominus \check{B}) \oplus B = \bigcup \{B + x \mid (B + x) \subset A\}$
Closure of A by B	$AfB = (A \oplus \check{B}) \ominus B = \bigcap \{B + x \mid (B + x) \cap A \neq \varnothing\}$

(See Figure 9.1.)
Clearly, $A \cup B$, $A \cap B$, A^c, and $\check{A}$ are easily realized by logical and, or, and negation on the image analyzer output, and by rotating the array. $A \oplus B$ can be found from

$$A \oplus B = \bigcup_{y \in B} (A + y) \qquad (9.2)$$

So we score one for $A \oplus B$ whenever a one occurs in any of the translated arrays $A + y$. Serra introduced this calculus for sets; his texture analyzer (see Klein and Serra, 1972) used a triangular grid which made calculation of $A \oplus B$ for a hexagon B easy; hexagons were used as approximations to circles.

We will use "$A \cap B \neq \varnothing$" sufficiently often to introduce the shorthand "$A \uparrow B$" read "A hits B." With our indicator-function theory we can find probabilities of events of the type

$$\{K \subset A, \quad K' \uparrow A\} \qquad (9.3)$$

Clearly, a stationary point process is a random set A, but we would expect the probability of (9.3) to be zero. Since point processes are useful starting points from which to build random sets by Serra's calculus, we need more general sets K and K' in (9.3). To do so, we must restrict the class of random sets; we assume that A is closed. This amounts to allocating the grain-pore boundary surface to the grains. Our basic events to which we assign probabilities then become

$$\{A \uparrow T_i, \quad i = 1, \ldots, n, \quad A \cap T_0 = \varnothing\} \qquad (9.4)$$

for $T_i \in \mathfrak{T}$, a class of "test sets" that replace the points in (9.3). We will

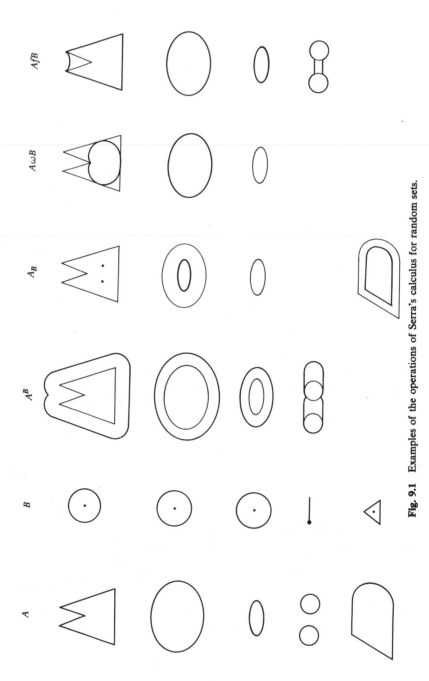

Fig. 9.1 Examples of the operations of Serra's calculus for random sets.

194

assume that finite unions of test sets are again test sets.   By the inclusion-exclusion formula we have

$$P(A \cap T_i = \varnothing, i = 1, \ldots, m, A \uparrow T_j, j = m+1, \ldots, n)$$

$$= \Sigma(-1)^{|\psi|-m} P(A \cap T = \varnothing \ \forall T \in \psi) \tag{9.5}$$

the sum being over all subcollections ψ of $(T_1, \ldots, T_n)$ which include $(T_1, \ldots, T_m)$.   Thus we need to specify

$$Q(T) = P(A \cap T = \varnothing) \qquad \forall T \in \mathcal{T}$$

So far we have avoided specifying $\mathcal{T}$.   Christensen (1974) explored equivalences in the choices of $\mathcal{T}$ and Kendall (1974) showed certain properties were crucial.   We have

$$T_i \nearrow T \text{ implies } Q(T_i) \searrow Q(T) \tag{9.6}$$

$$K_i \searrow K \text{ implies } Q(K_i) \nearrow Q(K) \tag{9.7}$$

$$Q(G) = \inf\{Q(K) | K \subset G\} \tag{9.8}$$

$$Q(K) = \sup\{Q(G) | G \supset K\} \tag{9.9}$$

where test sets denoted G and K are open and compact respectively.   We will start with $\mathcal{T}$ either a base for the topology of K, say all finite unions of open balls, or all compact sets.   Then Christensen's results show that $Q(T)$ is defined via (9.6)–(9.9) precisely for those sets which are countable unions of compact sets.

Theorem 9.1   (Choquet, Matheron, Kendall)

Q is $P(A \cap T = \varnothing)$ for some random closed set A if and only if

   (i) $Q(\varnothing) = 1$
  (ii) $T_1 \subset T_2$ implies $Q(T_1) \geqslant Q(T_2)$
 (iii) $\Sigma(-1)^{|\psi|} Q(\cup \{T_i | T_i \in \psi\}) \geqslant 0$ summed over all finite subcollections
       ψ of $(T_1, \ldots, T_n)$, any finite subset of $\mathcal{T}$.
 (iv) If $\mathcal{T}$ is a base
       $Q(T) = \inf\{Q(G) | G \subset K \subset T \text{ for some compact } K\}$.
       If $\mathcal{T}$ is the compact sets $K_n \searrow K$ implies $Q(K_n) \nearrow Q(K)$.

The reader may recognize from (9.5) that the expression in (iii) is just $P(A{\uparrow}T_i, i=1,\ldots,n)$. It can be checked (Matheron, 1975, Section 1.5) that $A\oplus K$, $\check{A}$, $A\ominus K$, $A\omega K$, AfK are all random sets for a compact set K.

In moving from finite sets to more general test sets we have lost something. For a single point set $\{x\}$,

$$\{\{x\}\subset A\}=\{\{x\}{\uparrow}A\}$$

so we could consider $\{K\subset A\}$ for *finite* sets K. Now $P(B\subset A)$ *is* defined for any subset B of X (since A is closed we need only check $A\supset B\cap D$ for a countable dense set D), but it may not be computable from Q.

Our stochastic process will be labeled by $\mathfrak{I}$ and give $I(A{\uparrow}T)$ for each $T\in\mathfrak{I}$. Whereas every closed set A gives a family of indicator functions the converse is no longer true, the conditions of Theorem 9.1 being those necessary to ensure that probability one is given to those realizations that do correspond to closed sets. Similar problems arise in the corresponding theorem for point processes (Ripley, 1976b), but we avoided that theory by constructing all our point processes via the Poisson process. We can also avoid Theorem 9.1, for our examples of random sets are constructed from Poisson processes.

Matheron (1967) defined a "Boolean scheme" as a generalization of a Poisson cluster process. Take a Poisson process with mean measure Λ and at each point form a disk or sphere with radius drawn independently from a specified distribution. The random closed set A is the union of the disks (or spheres). More generally we can replace the disk by an independent copy of a compact random set G, and let $A=\bigcup(G_i+\xi_i)$, where ξ_i are the points of the Poisson process and (G_i) independent copies of G. Suppose we look within a compact set D.

$$P\big((G_i+\xi_i)\cap T=\varnothing\,|\,\xi_i\in D\big)=\int_D Q_G(T-\xi_i)\,d\Lambda(\xi)/\Lambda(D)=\beta_D \text{ say}$$

$$P(A\cap T\cap D=\varnothing)$$

$$\approx P\big(\text{no }(G_i+\xi_i){\uparrow}T\text{ for }\xi_i\in D\big)$$

$$=\sum_n e^{-\Lambda(D)}(\beta_D)^n/n!=\exp\big[-\Lambda(D)(1-\beta_D)\big]$$

$$=\exp-\int_D\{1-Q_G(T-\xi)\}\,d\Lambda(\xi)$$

the approximation coming from discarding grains $G_i + \xi_i$ with centers outside D. If we let D increase to X in a suitably regular way

$$Q(T) = \exp\left[-\int_X \{1 - Q_G(T - \xi)\} \, d\Lambda(\xi) \right] \qquad (9.10)$$

We would like to have $P(T \subset A)$, but this seems impossible to compute in any useful form. This heuristic derivation of $Q(T)$ follows Watson (1975). Matheron (1975) gives a rigorous treatment and abstract characterizations of this and related processes.

The "Boolean scheme" is an unrealistic model for a rock, for no interaction between the grains is assumed. Other models can be constructed based on more general point processes, again avoiding Theorem 9.1. An example is to take the models for the centers of nonoverlapping disks proposed in Section 8.4 and to use the disks as a random set.

Another class of models is based on the random division of space. Pielou (1977, Chapter 12) describes two models for mosaics of vegetation. The cells of the division are filled independently with one of the possible types of vegetation. Models for the random division of space are surveyed by Miles (1972b); Pielou used two, the Dirichlet cells of a Poisson process and the tessellation created by a Poisson process of lines. The latter gives a simple sampling theory for a line intercept (Pielou, 1964, 1965; Switzer, 1965).

The models available are unsatisfactory and inferential techniques for them almost non-existent. Dupač (1980) gives a method to unravel the size distribution and mean number of circular grains of variable radius in a Boolean scheme.

Although we may not know how best to use them, estimates of quantities such as $P(T \subset A)$ are available for *homogeneous* random sets. An image analyzer views the random set A within a window D. Let ν denote area in the plane. Then by counting points on the grid we can estimate very accurately $\nu(C \cap D)/\nu(D)$, where C is a set created from A using Serra's calculus. This suggests estimating $P(T \subset A)$ by

$$\nu(\{x \mid T + x \subset A, x \in D\})/\nu(D) \qquad (9.11)$$

Unbiasedness of (9.11) follows by interchanging expectation and integration over D. The estimator of (9.11) is not practicable since $T + x$ might cross the boundary of D. To remove these edge problems we use a border (in the sense of Section 8.1) and allow x to vary in $D \ominus \check{T}$. Thus our unbiased estimator of $P(T \subset A)$ is

$$\frac{\#\{\mathbf{x}|T+\mathbf{x} \subset A, T+\mathbf{x} \subset D\}}{\#\{\mathbf{x}|T+\mathbf{x} \subset D\}}$$

To estimate $Q(T)$ we have

$$\frac{\#\{\mathbf{x}|T+\mathbf{x} \subset A^c, T+\mathbf{x} \subset D\}}{\#\{\mathbf{x}|T+\mathbf{x} \subset D\}} \tag{9.12}$$

This is the form used for $1 - \hat{p}(t)$ in Chapter 8 and is again unbiased.

There is a general procedure here. We have formed two new random sets $A' = A \ominus \check{T}$ and $A \oplus \check{T}$ and computed

$$\nu(A' \cap E)/\nu(E) \tag{9.13}$$

for $E = D \ominus \check{T}$, to estimate $P(T \subset A)$ and $1 - Q(T)$. Because A' is also stationary

$$E(\nu(A' \cap E)/\nu(E)) = \int_E P(\mathbf{x} \in A') \, d\mathbf{x}/\nu(E) = P(\mathbf{0} \in A')$$

which proves unbiasedness of (9.13) for $P(\mathbf{0} \in A')$ for any random set A' that can be constructed within E from $A \cap D$ using Serra's operations.

The variance of (9.13) is given by

$$E\left(\nu(A' \cap E)^2/\nu(E)^2\right) - P(\mathbf{0} \in A')^2$$

$$= \int_{E \times E} \{P(\mathbf{x} \in A', \mathbf{y} \in A') - P(\mathbf{x} \in A')P(\mathbf{y} \in A')\} \, d\mathbf{x} \, d\mathbf{y}/\nu(E)^2$$

$$= \int_{E \times E} \text{cov}(I_{A'}(\mathbf{x}), I_{A'}(\mathbf{y})) \, d\mathbf{x} \, d\mathbf{y}/\nu(E)^2 \tag{9.14}$$

Now $\text{cov}(I_{A'}(\mathbf{x}), I_{A'}(\mathbf{y}))$ is just the covariance function $C(\mathbf{y} - \mathbf{x})$ of our zero–one stochastic process Z abandoned earlier. However

$$C(\mathbf{h}) = P(\{\mathbf{0}, \mathbf{h}\} \subset A) - P(\mathbf{0} \in A)^2$$

$$= Q(\{\mathbf{0}, \mathbf{h}\}) - Q(\{\mathbf{0}\})^2$$

and so can be estimated itself from (9.12) and the estimate (possibly smoothed or fitted by a parametric model) used in (9.14).

9.2 BASIC QUANTITIES

Suppose we have a compact convex body Y on which we wish to make measurements. We will work in d-dimensional space ($d=2$ or 3) although Y might be of lower dimensionality, for instance, a surface or curve in $\mathbb{R}^3$. Let μ_k denote k-dimensional content (count, length, area, or volume). Define

$$\psi_0(Y) = \mu_d(Y)$$

$$\psi_k(Y) = E\{\mu_{d-k}(\Pi_S Y)\} \qquad 0 < k < d \qquad (9.15)$$

where S is a uniformly distributed k-dimensional direction and Π_S denotes the projection of Y onto the $(d-k)$ dimensional subspace determined by S. Then the *Minkowski functionals* are defined by

$$W_k^{(d)}(Y) = b_d \psi_k(Y)/b_{d-k} \qquad 0 \leqslant k < d$$

$$W_d^{(d)}(Y) = b_d \qquad (9.16)$$

where

$$b_k = \pi^{k/2}/\Gamma(1 + k/2)$$

is the content of the unit ball in $\mathbb{R}^k$. For a ball Y of radius R

$$W_k^{(d)}(Y) = b_d R^{d-k} \qquad (9.17)$$

These quantities can be expressed in simple cases in more familiar terms. $F(Y) = dW_1^{(d)}(Y)$ is the surface area (or perimeter length) of Y. $N(Y) = dW_{d-1}^{(d)}(Y)$ is the average over all orientations of the *caliper diameter* in that direction, the distance between two parallel planes (or lines) touching the body. If the body has a smooth enough surface ∂Y to define curvatures then there are $(d-1)$ principal curvatures $\kappa_1, \ldots, \kappa_{d-1}$. (The surface can be thought of as locally quadratic in a coordinate system with origin at the point of consideration. The κ_i are the eigenvalues of this quadratic form.) In $\mathbb{R}^2$ κ_1 is the inverse of the (local) radius of curvature, whereas in $\mathbb{R}^3$ κ_1 and κ_2 are the inverse of the radii of curvature of the most curved and least curved lines on the surface through the point being considered. $M = dW_2^{(d)}$ is the integral over the surface of the average of these curvatures (Miles, 1975; Santaló, 1976, p. 222). Thus we have in the practical cases

$\mathbb{R}^1$ $W_0(Y) = L(Y)$ length

        $W_1(Y) = P(\partial Y) = 2$ number of end points

$\mathbb{R}^2$ $W_0(Y) = A(Y)$ area

        $F = N = 2W_1(Y) = L(\partial Y)$ perimeter length

        $M = 2W_2(Y) = C(\partial Y) = 2\pi$ total curvature

$\mathbb{R}^3$ $W_0(Y) = V(Y)$ volume

        $F = 3W_1(Y) = S(\partial Y)$ surface area

$$N = M = 3W_2(Y) = K(\partial Y) = \frac{1}{2}\int_{\partial Y}(\kappa_1 + \kappa_2) \quad \text{integral mean curvature}$$

$$3W_3(Y) = G(\partial Y) = \int_{\partial Y}\kappa_1\kappa_2 = 4\pi \qquad \text{Gaussian curvature}$$

Minkowski functionals are important because $W_k^{(d)}$ is $(d-k)$ "dimensional," they are invariant under rigid motions, increasing, continuous (for a suitable topology on compact sets) and additive in the sense

$$\psi(X \cap Y) + \psi(X \cup Y) = \psi(X) + \psi(Y) \tag{9.18}$$

whenever $X, Y, X \cup Y$ are convex. Furthermore, any functional with these properties is a linear combination of Minkowski functionals (Hadwiger, 1957, pp. 221–225).

Clearly the intuitive interpretations can be extended to nonconvex bodies. The additivity allows us to extend the definitions to finite unions of disjoint convex sets, and in fact we can go to *ovoids*, arbitrary finite unions of compact convex sets, by defining

$$W_r(Y) = \sum_i W_r(Y_i) - \sum_{i<j} W_r(Y_i \cap Y_j) \cdots + (-1)^{m-1} W_r(Y_1 \cap \cdots \cap Y_m)$$

$$\tag{9.19}$$

where $Y = Y_1 \cap \cdots \cap Y_m$ is any decomposition of an ovoid into compact convex sets. The intuitive interpretations hold true for many compact sets, $C(\partial Y)$ being 2π times the number of bodies *minus* the number of holes and $G(\partial Y)$ 4π times the number of bodies and holes. Figure 9.2 gives an intuitive interpretation. Note that the original definition (9.15) holds only for convex bodies.

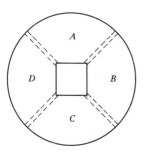

Fig. 9.2 The body shown can be divided into four convex bodies A, B, C, D with the double boundaries being the overlap. From (9.19) the total curvature is $4 \times 2\pi$ for A, B, C, D minus $4 \times 2\pi$ for the four dashed convex bodies, hence zero.

Image Analyzer Measurements

Watson (1975) shows how an image analyzer may be used to compute Minkowski functionals of a planar set Y within D. Obviously we can find the area $A(Y \cap D)$ by counting ones in the binary representation, and we will relate all the other measurements to areas of sets transformed by the operations of Serra's calculus. If Y has a smooth boundary and B is a ball of small enough radius r

$$A(Y \oplus B) = A(Y) + rL(\partial Y) + \tfrac{1}{2} r^2 C(\partial Y) \tag{9.20}$$

(this is intuitive and is proved by Miles, 1974b, Theorem 3). So if we measure $A(Y \oplus B)$ for different radii and plot this against radius we can estimate $L(\partial Y)$ and $C(\partial Y)$ by fitting a quadratic to the curve. Forming $Y \oplus B$ involves translating by all points in B, so can be expensive. Suppose we just use $T = \{0, \mathbf{h}\}$. Let $K(\mathbf{h}) = A(Y \ominus \check{T}) = A(Y \cap (Y - \mathbf{h}))$. If $\mathbf{h} = r\boldsymbol{\alpha}$ for a direction $\boldsymbol{\alpha}$, scalar $r > 0$ then

$$D_\alpha = -\left[\frac{d}{dr} K(r\boldsymbol{\alpha}) \right]\bigg|_{r=0} = \tfrac{1}{2} \int_{\partial Y} |\mathbf{n} \cdot \boldsymbol{\alpha}| \, ds \tag{9.21}$$

where $\mathbf{n}$ is the outward normal at the surface. (A few sketches will convince you that $2K(0) - 2K(r\boldsymbol{\alpha}) \approx r \int_{\partial A} \inf(0, \mathbf{n} \cdot \boldsymbol{\alpha}) \, ds$. $\int_{\partial A} \mathbf{n} \cdot \boldsymbol{\alpha} \, ds = 0$ because the boundary is one or more closed curves). The analyzer can easily form an approximation to D_α. However, from (9.21)

$$\operatorname*{ave}_\alpha D_\alpha = L(\partial A) \left\{ \tfrac{1}{2} \operatorname*{ave}_\alpha |\mathbf{n} \cdot \boldsymbol{\alpha}| \right\} = L(\partial A)/\pi \tag{9.22}$$

so by averaging over directions we can estimate the perimeter length, and if $\mathbf{y}$ is the line segment from $\mathbf{0}$ to $r\boldsymbol{\beta}$, $\boldsymbol{\beta}$ perpendicular to $\boldsymbol{\alpha}$,

$$D_\alpha(Y \oplus \mathbf{y}) - D_\alpha(Y) \approx r C(\partial Y) \tag{9.23}$$

This measurement is simple with a rectangular lattice binary representation of Y.

Counting objects

It is far more difficult than appears at first sight to devise an unbiased method for counting the number of objects within a window D. What should the observer count for objects that meet the boundary of D? The obvious suggestion to estimate the proportion of the object within D only works if the observer can see beyond the window, and is time consuming.

The net tangent count illustrated in Figure 9.3 provides a simple way to manually measure $C(\partial Y)$ and hence the number of bodies if these have no holes. It works well even if we can only observe $\partial Y \cap D$, for if a large domain is partitioned into subregions, the sum of the counts in the subregions will be the total count, so there is no systematic edge-effect bias. DeHoff (1978) gives an expository account of the tangent count method. An image analyzer can recognize tangent counts as shown in Figure 9.3.

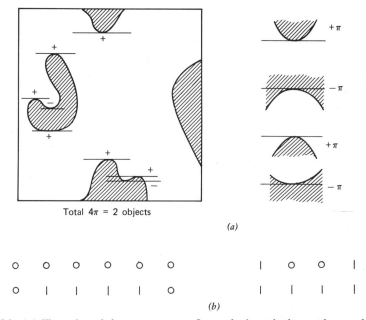

Total 4π = 2 objects

(a)

(b)

Fig. 9.3 (*a*) Illustration of the tangent count. Locate horizontal edges and score the four cases as shown. (*b*) Image analyser patterns (of arbitrary width) corresponding to tangent detection.

Figure 9.4 illustrates the *A*ssociated *P*oint and tiling methods. The *AP* method defines a unique point on each object and includes an object in the count when its associated point is within the window. An *AP* can be defined automatically by the illustrated "lower left" rule. This is precisely the tiling method of Gundersen (1977) for rectangular tiles. Miles (1978a) states that at least one image analyzer can find associated points as a standard function. It is not possible to find an *AP* from that part of an object within *D*.

Miles (1974b) describes two methods based on weighting objects. For "minus sampling" he considers only those objects totally contained in *D*. The idea is similar to that used for $\hat{K}$ in Section 8.3. For each object O_i contained in *D* compute a measurement T_i and

$$M_i = \text{area}\{x | x + O_i \subset D\} = \text{area}(D \ominus \breve{O}_i)$$

so M_i measures the freedom of O_i to move in *D*. Then

$$E(\Sigma T_i / M_i) = \rho P(M > 0) E(T | M > 0) \qquad (9.24)$$

where ρ is the number of objects per unit area, and the conditioning arises because large objects will never be measured. For counting $T_i = 1$, so $\Sigma 1 / M_i$ is an unbiased estimator of the intensity of objects that will fit

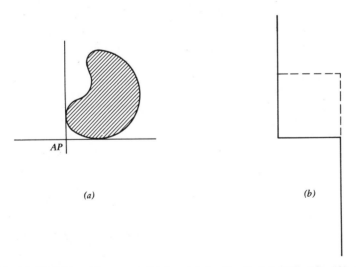

(a)   (b)

Fig. 9.4 (*a*) Definition of an associated point by the "lower left" rule. (*b*) One of Gundersen's tiling patterns. Objects are counted which have some point within the rectangle and do not meet the solid line.

within D.   Plus sampling is similar; all objects that meet D are included in the summation, and the weight is $M_i = \text{area}(D \oplus O_i)$.   Miles computes M for various objects within a *circular* domain D.   This method is only practical for image analyzers.   Even these have to recognize distinct objects before computing $D \ominus \breve{O}_i$ or $D \oplus O_i$.

Gundersen (1978) describes an elaborate tangent and edge count method of Giger (1967) and Weibel (1973).

9.3   STEREOLOGICAL SAMPLING

The purpose of this section is to investigate estimators of $V(Y)$, $S(\partial Y)$, $K(\partial Y)$, and $G(\partial Y)$ for a phase Y embedded in a specimen X; we assume X to be convex.   We assume that the content of X can be measured exactly. We denote by T and L a planar section and a linear probe respectively and allow only measurements in the section or the line.   Then the available estimators are given in Table 9.1.   Some of these formulas go back to Delesse (1848).   The ratio form of these estimates is intuitively sensible and will be justified by random sampling.   To be specific we mainly consider planar sampling of a three-dimensional body.

There is no such object as a uniformly distributed plane in $\mathbb{R}^3$, but there is an invariant *measure* for planes.   We denote by $H(X)$ the measure of the set of planes which hit X.   We can construct a plane (called an *IUR* plane, for isotropic uniform random) *conditional* on hitting X with probability proportional to this measure by taking a ball containing X, and generating an independent uniformly distributed normal ϕ and distance p

Table 9.1   Stereological Relationships

	Dimension			
Of X	Of Section	Of Y	Estimate	Of
3	2	3	$A(Y \cap T)/A(X \cap T)$	$V(Y)/V(X)$
3	2	2	$L(Y \cap T)/A(X \cap T)$	$(\pi/4)S(Y)/V(X)$
3	2	2	$C(Y \cap T)/A(X \cap T)$	$K(Y)/V(X)$
3	2	1	$P(Y \cap T)/A(X \cap T)$	$\frac{1}{2}L(Y)/V(X)$
3	1	3	$L(Y \cap L)/L(X \cap L)$	$V(Y)/V(X)$
3	1	2	$P(Y \cap L)/L(X \cap L)$	$\frac{1}{2}S(Y)/V(X)$
2	1	2	$L(Y \cap L)/L(X \cap L)$	$A(Y)/A(X)$
2	1	1	$P(Y \cap L)/L(X \cap L)$	$(2/\pi)L(Y)/A(X)$

from the center of the ball. These determine an *IUR* plane conditional on hitting the ball. If we reject those realizations which miss X we have an *IUR* plane T through X. Then

$$E\{A(Y \cap T)\} = V(Y)/H(X)$$

$$E\{L(Y \cap T)\} = (\pi/4)S(Y)/H(X) \qquad (9.25)$$

$$E\{C(Y \cap T)\} = K(Y)/H(X)$$

$$E\{P(Y \cap T)\} = \tfrac{1}{2}L(Y)/H(X)$$

From the description of the *IUR* plane it is clear that a typical left hand side is, for convex Y,

$$E\left\{ \int W_k^{(2)}(Y \cap S_s)\, ds \right\} / H(X) \qquad (9.26)$$

where the expectation is over a uniform direction S and S_s is the plane with normal S and distance s from the origin of the ball. Now (9.26) is an increasing additive functional invariant under rigid motions and of "dimension" $3 - k$, so by Hadwiger's theorem it is a multiple of $W_k^{(3)}(Y)$. By taking Y as a unit ball, we can find the multiple. Extending the argument to e-dimensional sections of d-dimensional bodies we have

$$E\left(W_k^{(e)}(Y \cap T) \right) = \left\{ W_k^{(d)}(Y)/H(X) \right\} \left\{ \frac{b_e}{b_d} \frac{b_{d-k}}{b_{e-k}} \right\} \qquad (9.27)$$

This holds for convex Y; we can extend by additivity to ovoids. (This argument follows Matheron, 1975, p. 82. An alternative derivation avoiding Hadwiger's theorem is given by Miles and Davy, 1976; Davy and Miles, 1977.)

From (9.25) and (9.27) we have that all the results of Table 9.1 satisfy $E(A)/E(B) = C$ when A/B estimates C. However, the expectation of the ratio is not the ratio of the expectations. If m sections are taken $(A_1 + \cdots + A_m)/(B_1 + \cdots + B_m)$ will be a reasonably unbiased estimator, but the point of using ratios is that A_i and B_i will be highly correlated. The standard remedy is *weighted* samples, introduced in this context by Miles and Davy (1976), Davy and Miles (1977) and Miles (1978b). If the probability density of a section is proportional to B, then

$$E_B(A/B) = \int \left(\frac{A}{B} \right) \frac{B}{E(B)}\, dP = E(A)/E(B) \qquad (9.28)$$

Thus the estimators of Table 9.1 will be unbiased if we weight planes by
$A(X \cap T)$ and lines by $L(X \cap L)$. Clearly, these are sensible weightings.
In fact, generating weighted planes is as easy as generating IUR planes.
Choose a random point within X and an independent random direction,
then use the plane through the point perpendicular to the chosen direction.
To find a length-weighted line, choose an area-weighted plane, take a
random point in this plane and a line with an independent uniform
direction through that point. Davy and Miles (1977) prove formally that
these algorithms do yield the desired distributions of planes and lines.

Note that these results apply only to the randomization used in sampling, and not to any inference from the specimen X to its population.
Furthermore, no variance results are given, although Miles and Davy
(1976) do repeat a standard argument suggesting that weighted samples
will have a small variance. In practice, random sampling is probably
impossible. Unbiasedness can be proven under other conditions. If the
specimen X is a cuboid it could be sectioned at a random height parallel to
one of the faces, so $A(X \cap T)$ is constant. However, we cannot use (9.25)
(except the first formula), since sampling is not isotropic, *unless* we assume
that the specimen was chosen isotropically from a population. Again, the
estimators of Table 9.1 will be unbiased. These assumptions seem closer
to the procedures used in practice. Serial sections are commonly taken
from a cuboid. Estimates from each are then unbiased but correlated.
So the elegant mathematics that results from random sampling is unsatisfactory both statistically and in practice.

There are two important omissions from Table 9.1. We have no
estimators of $G(\partial Y)$ from planar sections, nor of $C(\partial Y)$ from linear probes
of planar objects. These are important, for they are essentially counts of
the number of components of the Y phase. Miles and Davy (1978) have
found an unbiased estimator of $G(\partial Y)$ from measurements at the edge of a
wedge-shaped section, but this does not seem practical.

Practical problems that need further attention include the use of serial
sections and measurements other than the basic ones derived from
Minkowski functionals. The oil industry is interested in the topology and
many other properties of the pore space related to fluid flow, for example.
The new science of X-ray tomography (see Shepp and Kruskal, 1978 for a
mathematical report) may allow a three-dimensional digitized image to be
produced that allows a much greater selection of measurements and
eliminates the stereology completely!

9.4   SIZE DISTRIBUTIONS

Determining the grain sizes in a sedimentary rock is an important part of
the description of the rock. In practice, these are measured by comparing

a planar section with a set of standard charts whose number of grains per unit area is known. Specifically, the *ASTM* grain number is g, where 50×2^g is the number of grains per square inch, and charts are available for small integers g. Clearly, at this level of precision it is not necessary to define too carefully what is being measured. There are two problems in a more formal treatment. One is to measure the apparent grain sizes in a planar sample. This is considered under the heading of *granulometry*. The other is to infer the true distribution of grain sizes from that observed in a planar sample, known as *unfolding*. We ignore the problems of breaking grains during preparation of the sample, which can be either by cracking the rock or polishing a section, but do consider that the smallest grains might fail to be observed or be removed during the polishing process and the pits left go undetected.

Granulometry

Suppose we have a planar image of disjoint grains and pore space, to be analyzed by an image analyzer. Let A be the set of grains. Matheron (1967, 1975) abstracted the notion of sieving grains. He considered a family ψ_λ of transformations on A, where $\psi_\lambda(A)$ will be those grains too large to pass through the sieve of size λ. Clearly we would demand

1. $\psi_\lambda(A) \subset A$
2. $\lambda \geqslant \mu \Rightarrow \psi_\lambda(A) \subset \psi_\mu(A)$
3. $\psi_\mu(\psi_\lambda(A)) = \psi_{\max(\lambda, \mu)}(A)$
4. $B \subset C \subset A \Rightarrow \psi_\lambda(B) \subset \psi_\lambda(C)$
5. Equivariance under rigid motions T, so $\psi_\lambda(TA) = T\psi_\lambda(A)$

Fortunately we can achieve these without identifying the grains themselves. Consider $\psi_\lambda(A) = A\omega B_\lambda$, where ω is the opening defined in Section 9.1. We need B_λ to increase with λ and $B_\lambda = B_\lambda \omega B_\mu$ for $\lambda > \mu$. Typically, B_λ is a ball of diameter λ, in which case $\psi_\lambda(A)$ contains contributions only from those grain images with diameter (the greatest distance between two points in the grain) exceeding λ. Then a grain size distribution function can be defined by

$$G(\lambda) = 1 - \operatorname{area}(\psi_\lambda(A))/\operatorname{area}(A) \qquad (9.29)$$

If the grain images are circular G is just the empirical distribution function of the image diameters. If the image is viewed through a window, some edge images will be noncircular and a slight bias will be introduced; in particular, $G(\lambda)$ will no longer be a step function.

It is quicker for both automatic and manual measurements to take a *linear granulometry* in which B_λ is a line segment of length λ. For an image analyzer, this would be taken to be in the direction of the scan. Suppose A is scanned in a series of parallel traverses. Then the total traverse length across grains is proportional to area(A), and area($A \omega B_\lambda$) is proportional to the sum of all traverse lengths across grains which exceed λ. Thus $G(\lambda)$ is calculated from

$$G(\lambda) = \Sigma T_i I(T_i > \lambda) / \Sigma T_i$$

for grain traverse lengths (T_i). The larger grains will be measured on several traverses, so $G(\lambda)$ does not estimate the grain diameter distribution F, but

$$\int_0^\lambda s \, dF(s) \Big/ \int_0^\infty s \, dF(s)$$

Of course, F can be recovered from this weighted distribution.

Unfolding

Unfolding is the term used to describe the derivation of the distribution F (possibly with density f) of the particle sizes in $\mathbb{R}^3$ from the density g or distribution function G of their observed sizes in a planar section. To be specific, we will consider spherical particles the sizes of which are measured by their diameters. This problem was posed and solved analytically by Wicksell (1925). Suppose that the centers of the spheres are from a Poisson process of intensity λ, low enough for the spheres not to overlap. Then if m is the first moment of F,

$$g(y) = y \int_y^\infty (x^2 - y^2)^{-1/2} \, d\bar{F}(x) / m \qquad (9.30)$$

A random IUR sectioning plane will give the same formula for any collection of spheres in a compact body. It follows from two facts: spheres are cut with probabilities proportional to their diameters, and conditional on being cut the cut is uniformly distributed across the sphere. Thus, P(observed diameter $\geqslant y$|sphere of diameter x cut) = $p(y, x) = \sqrt{(x^2 - y^2)}/x$ and

$$g(y) = \int_y^\infty -\frac{d}{dy} p(y, x) x \, dF(x) / m$$

The inversion of this Abel-type integral equation is straightforward.

$$\int_x^\infty g(u)\frac{du}{(u^2-x^2)^{1/2}}$$

$$=\frac{1}{m}\int_x^\infty\left\{\int_x^v(v^2-y^2)^{-1/2}(y^2-x^2)^{-1/2}y\,dy\right\}dF(v)$$

and the term in { } is a constant, in fact $\pi/2$. Thus

$$1-F(x)=\frac{2m}{\pi}\int_x^\infty(u^2-x^2)^{-1/2}\,dG(u) \tag{9.31}$$

$$f(x)=-\frac{2mx}{\pi}\int_x^\infty(u^2-x^2)^{-1/2}\frac{d}{du}\left(u^{-1}g(u)\right)du \tag{9.32}$$

The constant m is the mean of F and so unknown; it is found from $F(0)=0$ or $\int_0^\infty f(x)\,dx=1$. We can relate the (noncentral) moments of F and G by

$$\mu_k(G)=\frac{\sqrt{\pi}}{2}\frac{\Gamma\{\frac{1}{2}(k+2)\}}{\Gamma\{\frac{1}{2}(k+3)\}}\frac{\mu_{k+1}(F)}{m} \tag{9.33}$$

Nicholson (1970) noted that we may not be able to observe small cross sections and proposed a truncation point y_0 for the diameters of the circles; smaller circles are ignored. Then (9.31) and (9.32) hold for $x\geqslant y_0$. However, there is now no information on $F(y_0-)$, the number of spheres of diameter less than y_0 and so m is unknown. We must use the conditional distribution for diameters not less than y_0.

A slightly different experimental setup is to view thin slices of particles embedded in a translucent material. Then what is observed is not the diameter of a cross section, but the maximum diameter within a slice of thickness t. We can amend (9.30) to

$$g(y)=\frac{1}{m+t}\left[y\int_y^\infty(x^2-y^2)^{-1/2}f(x)\,dx+tf(t)\right] \tag{9.34}$$

The inversion of (9.34) is more complicated, particularly if a truncation point is used, but has been considered by Bach (1967) and others.

The estimation of m is important, for we would expect to see λmA circles in a section of area A. The estimation of λ is often the main aim of the investigation, so we need an estimate $\hat{m}$ of m to find

$$\hat{\lambda} = \text{number of circles}/\hat{m}A \qquad (9.35)$$

Note that with no assumption on the particle shape we failed to estimate $G(\partial Y)/4\pi$, the number of particles, in Section 9.3. From (9.33)

$$m = \pi/2 \int_0^\infty u^{-1} dG(u) \qquad (9.36)$$

It should be clear from (9.35) and (9.36) that small diameters are important in finding m and $\hat{\lambda}$, yet these may be observed or measured or counted inaccurately. Suppose the observed circle diameters are $y_1, \ldots, y_n$. To use (9.31) or (9.36) we have to estimate G. The obvious choice is to take the empirical distribution function G_n of $\{y_1, \ldots, y_n\}$ and estimate m and F by

$$\hat{m} = \pi/\{2\Sigma 1/y_i\} \qquad (9.37)$$

$$\hat{F}(x) = 1 - \left\{ \sum_{y_i > x} (y_i^2 - x^2)^{-1/2} \right\} \Big/ \{\Sigma 1/y_i\} \qquad (9.38)$$

Watson (1971b) investigated these estimators. $\hat{F}$ is not monotone, decreasing to minus infinity at each y_i, and both estimators have infinite variance. The estimator $\hat{m}$ is asymptotically Normal, but at a rate that depends on the behavior of $F(x)$ for small x. Simulations showed the estimates to be useless.

The usual remedy for such behavior is to smooth G_n either nonparametrically or by fitting a parametric family for F or G. Anderssen and Jakeman (1975a, b) (using results from Jakeman and Anderssen, 1975) showed that the numerical inversion of (9.31) and (9.32) is unstable, as might be expected from Watson's results. Thus if a parametric family is used for G it is important that the inversion be performed analytically. Anderssen and Jakeman recommend using (9.31) with product integration, which is a means of interpolating G_n before integrating. They show this gives a monotone, consistent estimate of F (as $n \to \infty$). If the density f is required they recommend spectral differentiation of this estimate, which is again a smoothing technique.

Keiding et al. (1972) parametrized F as the distribution of the square root of a gamma variable with integer shape parameter, for which g can be found analytically from (9.30). They found maximum likelihood estimates numerically. In fact, they extended the parametric family by considering mixtures with three different scale factors, and assuming that glancing cuts on particles were missed. It was assumed that only cuts with ψ as shown in Figure 9.5 greater than ϕ were recorded. Then (9.30) becomes

$$g(y) = \frac{y}{m \cos \phi} \int_{y}^{y/\sin \phi} (x^2 - y^2)^{-1/2} \, dF(x)$$

They estimated a shape parameter, a scale parameter, ϕ, and the proportions in each of the three populations. Bimodal distributions F occur in practice so it seems important to take mixtures. Anderssen and Jakeman (1975b) give an example in which a restricted form of this parametric technique ($\phi = 0$, no mixtures) performs less well than does their own nonparametric scheme.

A parametric technique has advantages in the final stages of the analysis but some nonparametric estimate seems to be needed to choose the number of mixture components and give good guesses for the parameters to start the numerical maximization of the likelihood.

Surely the commonest way to smooth an empirical distribution function is to use a histogram. Often the observed diameters will have been grouped during the measuring process. We can regard (9.30) as a triangular system of linear equations between histogram estimates. If the bins are labeled 1 to h with midpoints a_i, upper limits b_i and widths w_i then we approximate (9.30) by

$$w_i g(a_i) = a_i \sum_{j > i} w_j f(a_j) / \left\{ m \left(b_j^2 - a_i^2 \right)^{1/2} \right\}. \tag{9.39}$$

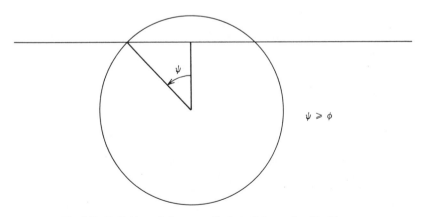

Fig. 9.5 Definition of the upper limit ϕ of the angle of incidence.

The upper limit b_j is used to avoid singularities. (Anderssen and Jakeman, 1975b discuss other methods in their Appendixes.) The procedure of Saltykov (1967) is essentially the solution of (9.39) by back substitution. All the observations in the largest size bin are assumed to come from spheres of diameter b_h; $f(a_h)$ is then computed to give the correct number in the largest bin and the contributions from those spheres to all the smaller diameter bins subtracted from the counts in those bins. This process is repeated for the second largest bin and so on. Eventually, the contents of some (small diameter) bin will become negative, and the densities for this bin and all smaller diameters are set to zero. Greeley and Crapo (1978) investigated the accuracy of this method for m by a simulation experiment. For 100–400 observed circles they recommended 10 bins as a compromise between accuracy and sampling variability. The method is simple and copes well with unobserved small circles. It could be developed by introducing criteria for amalgamating or splitting bins to give a good manual method of inversion. If the fine structure of f is important then Anderssen and Jakeman's method would be preferred.

So far we have only considered "size" as the diameter of a spherical particle. Wicksell (1926) considered ellipsoidal particles. Cruz Orive (1976, 1978) showed that one can unfold the distribution of a two-parameter family of ellipsoids of revolution but not the three-parameter family of general ellipsoids. The device originally suggested by Wicksell of regarding near-spherical particles and their cross sections as spheres and circles of the same volume and area is supported by the numerical experiments of Anderssen and Jakeman (1975) and Greeley and Crapo (1978). It seems sufficiently accurate in practice when the ratio of the longest axis to the shortest axis of an ellipsoidal particle does not exceed 2.

It is also possible to unfold the distribution of the length of line intercepts on a line probe. This is even more unstable, and the results of Nicholson (1970) strongly suggest that it should not be used whenever planar sectioning is possible.

A similar unfolding problem arises when measuring the distribution of membrane thickness, considered by Gundersen et al. (1978) and Jensen et al. (1979).

References

The compendia edited by Elias (1967), DeHoff and Rhines (1968), Nicholson (1972), Underwood et al. (1976), Chermant (1978), Miles and Serra (1978), and volume **95** of the *Journal of Microscopy* contain many papers on the practice and applications of stereology. Theoretical work is found in: Anderssen and Jakeman (1975a, b, 1976), Baldwin et al. (1971),

Cruz Orive (1976, 1978), Davy and Miles (1977), DeHoff (1962, 1967, 1978), Delfiner (1972), Delfiner et al. (1972), Giger (1967), Goldsmith (1967), Gundersen (1977, 1978), Gundersen et al. (1978), Jakeman and Anderssen (1975), Jensen et al. (1979), Keiding et al. (1972), Kendall and Moran (1963), Mayhew and Cruz Orive (1974), Miles (1972a, 1974b, 1975, 1978a, b), Miles and Davy (1976, 1977, 1978), Moran (1972), Nicholson (1970, 1976), Nicholson and Mercx (1969), Saltykov (1967), Santaló (1976), Solomon (1978), Tallis (1970), Watson (1971b, 1975), Weibel (1973), Weibel et al. (1972), Wicksell (1925, 1926).

Bibliography

Agterburg, F. P. (1967) Computer techniques in geology. *Earth Science Reviews* 3 47–77.

Agterburg, F. P. (1968) Application of trend analysis in the evaluation of Whaleback mine, Newfoundland. *Ore Reserve Estimation and Grade Control*. CIMM Special 9 77–96.

Agterburg, F. P. (1969) Interpolation of areally distributed data. *Quarterly of the Colorado School of Mines* 64 217–237.

Agterburg, F. P. (1970) Autocorrelation functions in geology. *Geostatistics* D. F. Merriam, Ed. Plenum, New York. 113–141.

Agterburg, F. P. and Chung, C. F. (1973) Geomathematical prediction of sulphur in coal, New Lingan mine area, Sydney coalfield. *Canadian Mining and Metallurgical Bulletin* 66(738) 85–96.

Aherne, W. A. and Diggle, P. J. (1978) The estimation of neuronal population density by a robust distance method. *Journal of Microscopy* 114 285–293.

Akima, H. (1974) A method of bivariate interpolation and smooth surface fitting based on local procedures. Algorithm 474. *Communications Association Computing Machinery* 17 18–20; 26–31.

Akima, H. (1978) A method of bivariate interpolation and smooth surface fitting for irregularly spaced data points. Algorithm 526. *ACM Transactions on Mathematical Software* 4 148–159; 160–164.

Alldredge, J. R. and Alldredge, N. G. (1978) Geostatistics: A bibliography. *International Statistical Review* 46 77–88.

Allen, T. F. (1973) Pattern analysis of an epiphyllous alga. *Journal of Ecology* 61 887–899.

Anderson, D. J. (1961a) The structure of some upland plant communities in Caernarvonshire. I. The pattern shown by *Pteridium aquilinum*. *Journal Ecology* 49 369–376.

Anderson, D. J. (1961b) II. *Vaccinium myrtillis* and *Calluna vulgaris*. *Journal of Ecology* 49 731–738.

Anderson, D. J. (1967) Studies on structure in plant communities. III. Data on pattern in colonizing species. *Journal of Ecology* 55 397–404.

Anderson, D. J. (1971) Spatial patterns in some Australian dryland plant communities. *Statistical Ecology*. G. P. Patil, E. C. Pielou, and W. E. Waters, Eds. Pennsylvania State University Press, College Town, Pa. I 271–286.

Anderson, D. R. and Pospahala, R. S. (1970) Correction of bias in belt transect studies of immotile objects. *Journal of Wildlife Management* 34 141–146.

214

Anderssen, R. S. and Jakeman, A. J. (1975a) Product integration for functionals of particle size distribution. *Utilitas Mathematica* **8** 111–126.

Anderssen, R. S. and Jakeman, A. J. (1975b) Abel type integral equations in stereology. II. Computational methods of solution and the random spheres approximation. *Journal of Microscopy* **105** 135–153.

Anderssen, R. S. and Jakeman, A. J. (1976) Computational methods in stereology. *National Bureau of Standards Special Publication* **431** 13–18.

Arora, S. S. and Brown, M. (1977) Alternatives to spatial autocorrelation: An improvement over current practice. *International Regional Science Review* **2** 67–78.

Arvantis, L. G. and O'Regan, W. G. (1972) Cluster or satellite sampling in forestry: A Monte Carlo computer simulation study. *IUFRO Third Conference Advisory Group of Forestry Statisticians*. Institut National de la Recherche Agronomique, Paris. 191–205.

Atkinson, A. C. (1979a) The computer generation of Poisson random variables. *Applied Statistics* **28** 29–35.

Atkinson, A. C. (1979b) Recent developments in the computer generation of Poisson random variables. *Applied Statistics* **28** 260–263.

Atkinson, R. J. A. (1974) Behavioural ecology of the mud-burrowing crab *Gonoplax rhomboides*. *Marine Biology* **25** 239–252.

Austin, M. P. (1968) Pattern in a *Zerna erecta* dominated community. *Journal of Ecology* **56** 197–218.

Bach, G. (1967) Size distribution of particles derived from the size distribution of their sections. *Stereology*. H. Elias, Ed. Springer, Berlin. 174–185.

Baddeley, A. (1980) A limit theorem for some statistics of spatial data. *Advances in Applied Probability* **12** 447–461.

Baker, R. J. and Nelder, J. A. (1977) *The GLIM System Manual*. Numerical Algorithms Group, Oxford.

Baldwin, J. P., Tinker, P. B. and Marriott, F. H. C. (1971) The measurement of length and distribution of onion roots in the field and the laboratory. *Journal of Applied Ecology* **8** 543–554.

Barnard, G. (1963) Contribution to the discussion of Bartlett. *Journal of the Royal Statistical Society* **B25** 294.

Barrett, J. R. (1965) Correction for edge effect bias in point sampling. *Forestry Science* **10** 52–55.

Bartels, C. P. A. (1979) Operational statistical methods for analysing spatial data. *Exploratory and Explanatory Statistical Analysis of Spatial Data*. C. P. A. Bartels and R. H. Ketellapper, Eds. Martinus Nijhoff, Boston, Mass. 5–50.

Bartels, C. P. A. and Ketellapper, R. H. (Eds.) (1979) *Exploratory and Explanatory Statistical Analysis of Spatial Data*. Martinus Nijhoff, Boston, Mass.

Bartholomew, D. J. (1973) *Stochastic Models for Social Processes*, 2nd ed. Wiley, Chichester.

Bartlett, M. S. (1938) The approximate recovery of information from field experiments with large blocks. *Journal of Agricultural Science* **28** 418–427.

Bartlett, M. S. (1963) The spectral analysis of point processes. *Journal of the Royal Statistical Society* **B25** 264–296.

Bartlett, M. S. (1964) The spectral analysis of two-dimensional point processes. *Biometrika* **51** 299–311.

Bartlett, M. S. (1971a) Two-dimensional nearest-neighbour systems and their ecological applications. *Statistical Ecology*. G. P. Patil, E. C. Pielou, and W. E. Waters, Eds. Pennsylvania State University Press, College Town, Pa. I 179–194.

Bartlett, M. S. (1971b) Physical nearest-neighbour models and non-linear time series. *Journal of Applied Probability* **8** 222–232.

Bartlett, M. S. (1972) Physical nearest-neighbour models and non-linear time series II. *Journal of Applied Probability* **9** 76–86.

Bartlett, M. S. (1974a) Physical nearest-neighbour models and non-linear time series III. *Journal of Applied Probability* **11** 715–725.

Bartlett, M. S. (1974b) The statistical analysis of spatial pattern. *Advances in Applied Probability* **6** 336–358.

Bartlett, M. S. (1975) *The Statistical Analysis of Spatial Pattern*. Chapman and Hall, London.

Bartlett, M. S. (1978a) Further analysis of spatial patterns: a re-examination of the Papadakis method of improving the accuracy of randomized block experiments. *Supplement Advances in Applied Probability* **10** 133–143.

Bartlett, M. S. (1978b) Nearest neighbour models in the analysis of field experiments. *Journal of the Royal Statistical Society* **B40** 147–174.

Bartlett, M. S. and Besag, J. (1969) Correlation properties of some nearest-neighbour systems. *Bulletin of the International Statistical Institute* **43(2)** 191–193.

Barton, D. E., David, F. N. and Fix, E. (1963) Random points in a circle and the analysis of chromosome patterns. *Biometrika* **50** 23–29.

Bary-Lenger, A. (1967) Étude statistique de la dispersion spatiale des arbres en fôret. *Biométrie Praximetrie* **8** 115–148.

Bassett, K. A. (1972) Numerical methods for map analysis. *Progress in Geography* **4** 217–254.

Batcha, J. P. and Reese, J. R. (1964) Surface determination and automatic contouring for mineral exploration, extraction and processing. *Quarterly of the Colorado School of Mines* **59(4)** 1–14.

Batcheler, C. L. (1971) Estimate of density from a sample of joint point and nearest-neighbour distances. *Ecology* **52** 703–709.

Batcheler, C. L. (1973) Estimating density and dispersion from truncated or unrestricted joint point-distance nearest-neighbour distances. *Proceedings of the New Zealand Ecological Society* **20** 131–147.

Batcheler, C. L. (1975) Probable limit of error of the point distance-neighbour distance estimate of density. *Proceedings of the New Zealand Ecological Society* **22** 28–33.

Batcheler, C. L. and Bell, D. J. (1970) Experiments in estimating density from joint point- and nearest-neighbour distance samples. *Proceedings of the New Zealand Ecological Society* **17** 111–117.

Batcheler, C. L. and Hodder, R. A. C. (1975) Tests of a distance technique for inventory of pine populations. *New Zealand Journal of Forestry Science* **5** 3–17.

Bauersachs, E. (1942) Bestandesmassenaufnahme nach dem Mittelstammverfahren des zweitkleinsten Stammabstandes. *Forstwissenschaften Cbl.* **64** 182–186.

Bengtsson, B. E. and Nordbeck, S. (1964) Construction of isarithms and isarithmic maps by computers. *BIT* **4** 87–105.

Bennett, R. J. (1975) Dynamic systems modelling of the North-West region. *Environment and Planning* **A7** 525–566; 617–636; 887–898.

Bennett, R. J. (1979) *Spatial Time Series*. Pion, London.

Berry, B. J. L. (1971) Problems of data organization and analytical methods in geography. *Journal of the American Statistical Association* **66** 510–523.

Berry, B. J. L. and Marble, D. F. (Eds.) (1968) *Spatial Analysis, A Reader in Statistical Geography*. Prentice-Hall, Englewood Cliffs, N.J.

Besag, J. E. (1972a) On the correlation structure of some two-dimensional stationary processes. *Biometrika* **59** 43–48.

Besag, J. E. (1972b) Nearest-neighbour systems and the auto-logistic model for binary data. *Journal of the Royal Statistical Society* **B34** 75–83.

Besag, J. E. (1972c) Nearest-neighbour systems: A lemma with application to Bartlett's global solutions. *Journal of Applied Probability* **9** 418–421.

Besag, J. (1974) Spatial interaction and the statistical analysis of lattice systems. *Journal of the Royal Statistical Society* **B36** 192–236.

Besag, J. (1975) Statistical analysis of non-lattice data. *The Statistician* **24** 179–195.

Besag, J. (1977a) Errors-in-variables estimation for Gaussian lattice schemes. *Journal of the Royal Statistical Society* **B39** 73–78.

Besag, J. (1977b) On spatial-temporal models and Markov fields. *Transactions of the Seventh Prague Conference*. 47–55.

Besag, J. (1977c) Efficiency of pseudo-likelihood estimators for simple Gaussian fields. *Biometrika* **64** 616–618.

Besag, J. (1978) Some methods of statistical analysis for spatial data. *Bulletin of the International Statistical Institute* **47(2)** 77–92.

Besag, J. and Diggle, P. J. (1977) Simple Monte Carlo tests for spatial pattern. *Applied Statistics* **26** 327–333.

Besag, J. E. and Gleaves, J. T. (1973) On the detection of spatial pattern in plant communities. *Bulletin of the International Statistical Institute* **45(1)** 153–158.

Besag, J. E. and Moran, P. A. P. (1975) On the estimation and testing of spatial interaction in Gaussian lattice processes. *Biometrika* **62** 555–562.

Bhattacharyya, B. K. (1965) Two-dimensional harmonic analysis as a tool for magnetic interpretation. *Geophysics* **30** 829–857.

Bhattacharyya, B. K. (1969) Bicubic spline interpolation as a method for treatment of potential field data. *Geophysics* **34** 402–423.

Bhattacharyya, B. K. (1971) An automatic method of compilation and mapping of high-resolution aeromagnetic data. *Geophysics* **36** 695–716.

Billingsley, P. (1961) *Statistical Inferences for Markov Processes*. University of Chicago Press, Chicago.

Bitterlich, W. (1948) Die Winkelzählprobe. *Allgemeine forst-und holzwirtschaftliche Zeitung.* **59** 4–5.

Blackith, R. E. (1958) Nearest-neighbour measurements for the estimation of animal populations. *Ecology* **39** 147–150.

Blackith, R. E., Siddorn, J. W., Waloff, N., and van Emden, H. F. (1963) Mound nests of the yellow ant *Lasius flavus* L. on waterlogged pasture in Devonshire. *Entomologist's Monthly Magazine* **99** 48–49.

Blais, R. A. and Carlier, P. A. (1967) Applications of geostatistics in ore reserve evaluation. *Ore Reserve Estimation and Grade Control*. CIMM Special volume. **9**, 41–68.

Bloomfield, P. (1976) *Fourier Analysis of Time Series: An Introduction*. Wiley, New York.

Bodson, D. and Peeters, D. (1975) Estimation of the coefficients of a linear regression in the presence of spatial autocorrelation: An application to a Belgian labour-demand function. *Environment and Planning* **A7** 455–472.

Bookstein, F. L. (1978) *The Measurement of Biological Shape and Shape Change*. Lecture Notes in Biomathematics **24** 1–191.

Boots, B. N. (1973) Some models of the random subdivision of space. *Geografiska Annaler* **55B** 34–48.

Boots, B. N. (1974) Delaunay triangles: An alternative approach to point pattern analysis. *Proceedings of the Association of American Geographers* **6** 26–29.

Boots, B. N. (1975) Patterns of urban settlements revisited. *The Professional Geographer* **27** 426–431.

Borel, E. (1925) *Principes et Formules Classiques du Calcul des Probabilités*. Gauther-Villars, Paris.

Brandsma, A. S. and Ketellapper, R. H. (1979) Further evidence on alternative procedures for testing of spatial autocorrelation amongst regression disturbances. *Exploratory and Explanatory Statistical Analysis of Spatial Data*. C. P. A. Bartels and R. H. Ketellapper, Eds. Martinus Nijhoff, Boston, Mass. 113–136.

Breiman, L. (1968) *Probability*. Addison-Wesley, Reading, Mass.

Brook, D. (1964) On the distinction between the conditional probability and the joint probability approaches in the specification of nearest-neighbour systems. *Biometrika* **51** 481–483.

Brown, D. (1975) A test of randomness of nest spacing. *Wildfowl* **26** 102–103.

Brown, D. and Rothery, P. (1978) Randomness and local regularity of points in a plane. *Biometrika* **65** 115–122.

Brown, G. S. (1965) Point density in stems per acre. *New Zealand Forestry Research Notes* **38** 1–11.

Brown, S. and Holgate, P. (1974) The thinned plantation. *Biometrika* **61** 253–261.

Brown, T. C. and Silverman, B. W. (1979) Rates of Poisson convergence for U-statistics. *Journal of Applied Probability* **16** 428–432.

Brush, J. E. (1953) The hierarchy of central places in Southwestern Wisconsin. *Geographical Review* **43** 380–402.

Burnham, K. P. and Anderson, D. R. (1976) Mathematical models for nonparametric inferences from line transect data. *Biometrics* **32** 325–336.

Byth, K. and Ripley, B. D. (1980) On sampling spatial patterns by distance methods. *Biometrics* **36** 279–284.

Campbell, D. J. and Clarke, D. J. (1971) Nearest neighbour tests of significance for non-randomness in the spatial distribution of singing crickets [*Teleogryllus commodus* (Walker)]. *Animal Behaviour* **19** 750–756.

Campbell, J. B. (1978) Join count autocorrelation for gridded data. *Computer Applications* **5** 893–920.

Carter, D. S. and Prenter, P. M. (1972) Exponential spaces and counting processes. *Zeitschrift für Wahrscheinlichkeitstheorie* **21** 1–19.

Cassetti, E. (1966) Analysis of spatial association by trigonometric polynomials. *Canadian Geographer* **10** 199–204.

Catana, A. J. (1963) The wandering quarter method of estimating population density. *Ecology* **44** 349–360.

Chambers, J. M. (1977) *Computational Methods for Data Analysis*. Wiley, New York.

Chayes, F. (1970) On deciding whether trend surfaces of progressively higher order are meaningful. *Bulletin of the Geological Society of America* **81** 1273–1278.

Chayes, F. (1972) Looking through rocks. *Statistics: A Guide to the Unknown*. J. M. Tanur, F. Mosteller, W. H. Kruskal, R. F. Link, R. S. Pieters, and G. R. Rising, Eds. Holden-Day, San Francisco, Calif. 362–371.

Chermant, J. L. (Ed.) (1978) Quantitative analysis of microstructures in materials science, biology and medicine. *Practical Metallography* **8** 1–455.

Chiles, J. R. and Matheron, G. (1975) Interpolation optimale et cartographie. *Annales des Mines* Nov. 19–26.

Chorley, R. J. and Haggett, P. (1965) Trend-surface mapping in geographical research. *Institute of British Geographers Transactions* **37** 47–67. Reprinted in *Spatial Analysis*. B. J. L. Berry and D. F. Marble, Eds. Prentice-Hall, Englewood Cliffs, N.J.

Christensen, J. P. R. (1974) *Topology and Borel Structure*. North-Holland, Amsterdam.

Clark, P. J. and Evans, F. C. (1954) Distance to nearest neighbour as a measure of spatial relationships in populations. *Ecology* **35** 445–453.

Clark, P. J. and Evans, F. C. (1955) On some aspects of spatial pattern in biological populations. *Science* **121** 397–398.

Cliff, A. D. and Kelly, F. P. (1977) Regional taxonomy using trend surface coefficients and invariants. *Environment and Planning* **A9** 945–955.

Cliff, A. D. and Ord, J. K. (1973) *Spatial Autocorrelation*. Pion, London.

Cliff, A. D. and Ord, J. K. (1975) Model building and the analysis of spatial pattern in human geography. *Journal of the Royal Statistical Society* **B37** 297–348.

Cliff, A. D., Haggett, P., Ord, J. K., Bassett, K., and Davies, R. B. (1975) *Elements of Spatial Structure: A Quantitative Approach*. Cambridge University Press, London.

Clifford, P. and Sudbury, A. (1973) A model for spatial conflict. *Biometrika* **60** 581–588.

Cole, A. J. (1970) A trend surface technique. *International Statistical Review* **38** 12–20.

Cooper, C. F. (1960) Changes in vegetation, structure and growth of Southwestern pine forests since white settlement. *Ecological Monographs* **30** 129–164.

Cooper, C. F. (1961) Pattern in Ponderosa pine forests. *Ecology* **42** 493–499.

Cormack, R. M. (1977) The invariance of Cox and Lewis's statistic for the analysis of spatial patterns. *Biometrika* **64** 143–144.

Cottam, G. (1947) A point method for making rapid surveys of woodlands. *Bulletin of the Ecological Society of America* **28** 60.

Cottam, G. and Curtis, J. T. (1949) A method for making rapid surveys of woodlands by means of pairs of randomly selected trees. *Ecology* **30** 101–104.

Cottam, G. and Curtis, J. T. (1956) The use of distance measures in phytosociological sampling. *Ecology* **37** 451–460.

Cottam, G., Curtis, J. T., and Catana, A. J. (1957) Some sampling characteristics of a series of aggregated populations. *Ecology* **38** 610–622.

Cowie, S. R. (1967) The cumulative frequency nearest-neighbour method for the identification of spatial patterns. *Department of Geography, University of Bristol Seminar Paper* **A10**.

Cox, D. R. and Lewis, P. A. W. (1966) *The Statistical Analysis of a Series of Events*. Methuen, London.

Cox, F. (1972) Are there any consistent parameters for distance methods if the spatial distribution deviates remarkably from a bidimensional Poisson process? *IUFRO Third Conference Advisory Group of Forestry Statisticians*. Institut National de la Recherche Agronomique, Paris. 247–259.

Cox, T. F. (1976) The robust estimation of the density of a forest stand using a new conditioned distance method. *Biometrika* **63** 493–499.

Cox, T. F. (1979) A method for mapping the dense and sparse regions of a forest stand. *Applied Statistics* **28** 14–19.

Cox. T. F. and Lewis, T. (1976) A conditioned distance ratio method for analysing spatial patterns. *Biometrika* **63** 483–491.

Craig, C. C. (1953) On a method of estimating biological populations in the field. *Biometrika* **40** 216–218.

Crain, I. K. (1970) Computer interpolation and contouring of two-dimensional data: A review. *Geoexploration* **8** 71–86.

Crain, I. K. and Bhattacharyya, B. K. (1967) Treatment of non-equispaced two-dimensional data with a digital computer. *Geoexploration* **5** 173–194.

Cramér, H and Leadbetter, M. R. (1967) *Stationary and Related Stochastic Processes: Sample Function Properties and Their Applications*. Wiley, New York.

Crisp, D. J. (1961) Territorial behaviour in barnacle settlement. *Journal of Experimental Biology* **38** 429–446.

Croxton, C. A. (1974) *Liquid State Physics—A Statistical Mechanical Introduction*. Cambridge University Press, London.

Cruz Orive, L.-M. (1976) Particle size–shape distributions: The general spheroid problem. I. Mathematical model. *Journal of Microscopy* **107** 235–253.

Cruz Orive, L.-M. (1978) II. Stochastic model and practical guide. *Journal of Microscopy* **112** 153–167.

Curry, L. (1964) The random spatial economy: An exploration in settlement theory. *Annals of the Association of American Geographers* **54** 138–146.

Curry, L. (1967) Central places in the random spatial economy. *Journal of Regional Science* **7** 217–238.

Dacey, M. F. (1962) Analysis of central place and point patterns by a nearest neighbour method. *Lund Studies in Geography* **B24** 55–75.

Dacey, M. F. (1963) Order neighbor statistics for a class of random patterns in multidimensional space. *Annals of the Association of American Geographers* **53** 505–515.

Dacey, M. F. (1965) The geometry of central place theory. *Geografiska Annaler* **47B** 111–124.

Dacey, M. F. (1968) A review of measures of contiguity for two and *k*-color maps. *Spatial Analysis*. B. J. L. Berry and D. F. Marble, Eds. Prentice-Hall, Englewood Cliffs, N.J. 479–495.

Dacey, M. F. and Tung, T. (1962) The identification of randomness in point patterns. *Journal of Regional Science* **4** 83–96.

Dagbert, M. and David, M. (1976) Universal kriging for ore-reserve estimation—Conceptual background and application to the Navan deposit. *Canadian Mining and Metallurgical Bulletin* **69**(766) 80–92.

Dalenius, T., Hajek, J. and Zubrzycki, S. (1961) On plane sampling and related geometrical problems. *Proceedings of the Fourth Berkeley Symposium on Mathematical Statistics and Probability*, **I** 125–150.

Daley, D. J. and Vere-Jones, D. (1972) A summary of the theory of point processes. *Stochastic Point Processes*. P. A. W. Lewis, Ed. Wiley, New York. 299–383.

David, F. N. and Barton D. E. (1966) Two space–time tests for epidemicity. *British Journal of Preventive and Social Medicine* **20** 44–48.

David, F. N. and Moore, P. G. (1954) Notes on contagious distributions in plant populations. *Annals of Botany* **18** 47–53.

David, M. (1977) *Geostatistical Ore Reserve Estimation*. Elsevier, Amsterdam.

David, M. (1978) Sampling and estimation problems for three dimensional spatial stationary and nonstationary stochastic processes as encountered in the mineral industry. *Journal of Statistical Planning and Inference* **2** 211–244.

Davis, J. C. (1973) *Statistics and Data Analysis in Geology*. Wiley, New York.

Davis, J. C. and McCullagh, M. J. (Eds.) (1975) *Display and Analysis of Spatial Data*. Wiley, New York.

Davis, J. C. and Preston, F. W. (1972) Optical processing: An alternative to digital computing. *Geological Society of America Special Paper* **146** 49–68.

Davis, M., Groth, E. J., and Peebles, P. J. E. (1977) Study of galaxy correlations: Evidence for the gravitational instability picture in a dense universe. *Astrophysical Journal* **212** L107–L111.

Davy, P. and Miles, R. E. (1977) Sampling theory for opaque spatial specimens. *Journal of the Royal Statistical Society* **B39** 56–65.

De Boor, C. (1962) Bicubic spline interpolation. *Journal of Mathematics and Physics* **41** 212–218.

DeHoff, R. T. (1962) The determination of the size distribution of ellipsoidal particles from measurements made on random plane sections. *Transactions of the Metallurgical Society of AIME* **224** 474–477.

DeHoff, R. T. (1967) The quantitative estimation of mean surface curvature. *Transactions of the Metallurgical Society of AIME* **239** 617–621.

DeHoff, R. T. (1978) Stereological uses of the area tangent count. *Lecture Notes in Biomathematics* **23** 99–113.

DeHoff, R. T. and Rhines, F. N. (Eds.) (1968) *Quantitative Metallography*. McGraw-Hill, New York.

Delesse, A. (1848) Procédé mécanique pour déterminer la composition des roches. *Annales Mines Belgiques* **13** 379–388.

Delfiner, P. (1972) A generalization of the concept of size. *Journal of Microscopy* **95** 203–216.

Delfiner, P. (1973) Analyse de géopotential et du vent géostrophique par krigeage universel. *Revue de la Métérologie* **25** 1–56.

Delfiner, P. (1976) Linear estimation of nonstationary spatial phenomena. *Advanced Geostatistics in the Mining Industry* M. Guarascio, C. J. Huijbregts, and M. David, Eds. Reidel, Dordrecht. 49–68.

Delfiner, P. and Delhomme, J. P. (1975) Optimal interpolation by kriging. *Display and Analysis of Spatial Data*. J. C. Davis and M. J. McCullagh, Eds. Wiley, New York. 96–114.

Delfiner, P., Étienne, J., and Fonck, J. M. (1972) Application de l'analyser de textures à l'étude morphologique des réseaux poreux en lames minces. *Revue de l'Institut Français du Pétrole et Annales de Combustibles Liquides* **XXVII** 535–558.

De Vos, S. (1973) The use of nearest neighbour distances. *Tijdschrift voor Economische en Sociale Geografie* **64** 307–319.

Diggle, P. J. (1975) Robust density estimation using distance methods. *Biometrika* **62** 39–48.

Diggle, P. J. (1976) Note on the Clark and Evans test of spatial randomness. Appendix to Hodder and Orton (1976), 246–248.

Diggle, P. J. (1977a) A note on robust density estimation for spatial point patterns. *Biometrika* **64** 91–95.

Diggle, P. J. (1977b) The detection of random heterogeneity in plant populations. *Biometrics* **33** 390–394.

Diggle, P. J. (1978) On parameter estimation for spatial point processes. *Journal of the Royal Statistical Society* **B40** 178–181.

Diggle, P. J. (1979) On parameter estimation and goodness-of-fit testing for spatial point patterns. *Biometrics* **35** 87–101.

Diggle, P. J. (1980) Statistical methods for spatial point patterns in ecology. *Spatial and Temporal Analysis in Ecology*. R. M. Cormack and J. K. Ord, Eds. International Co-operative Publishing House, Burtonsville, Md. 95–150.

Diggle, P. J. and Matérn, B. (1980) On sampling designs for the study of point-event nearest neighbour distributions in R^2. *Scandinavian Journal of Statistics* **7** 80–84.

Diggle, P. J., Besag, J., and Gleaves, J. T. (1976) Statistical analysis of spatial point patterns by means of distance methods. *Biometrics* **32** 659–667.

Donnelly, K. (1978a) Simulations to determine the variance and edge effect of total nearest neighbour distance. *Simulation Methods in Archaeology*. I. Hodder, Ed. Cambridge University Press, London.

Donnelly, K. P. (1978b) Total nearest neighbour distance: A statistic to test for clustering of points. Unpublished Essay.

Douglas, J. B (1975) Clustering and aggregation. *Sankhya* **37B** 398–417.

Douglas, J. B. (1979) *Analysis with Standard Contagious Distributions*. International Co-operative Publishing House, Burtonsville, Md.

Dupač, V. (1980) Parameter estimation in the Poisson field of discs. *Biometrika* **67** 187–190.

Durbin, J. (1971) Boundary-crossing probabilities for the Brownian motion and Poisson processes and techniques for computing the power of the Kolmogorov-Smirnov test. *Journal of Applied Probability* **8** 431–453.

Dutta, I. and Rao, S. V. L. N. (1977) Variograms and their role in the interpretation of soft iron ore types. *Mathematical Geology* **9** 99–111.

Eberhardt, L. L. (1967) Some developments in "distance sampling." *Biometrics* **23** 207–216.

Eberhardt, L. L. (1968) A preliminary appraisal of the line transect. *Journal of Wildlife Management* **32** 82–88.

Eberhardt, L. L. (1978) Transect methods for population studies. *Journal of Wildlife Management* **42** 1–31.

Eberhardt, L. L. (1979) Line transects based on right-angle distances. *Journal of Wildlife Management* **43** 768–774.

Edwards, A. W. F. (1972) *Likelihood*. Cambridge University Press, London.

Elias, H. (Ed.) (1967) *Stereology*. Springer-Verlag, Berlin.

Ellis, J. A., Westmeier, R. L., Thomas, K. P., and Norton, H. W. (1969) Spatial relationships among quail coveys. *Journal of Wildlife Management* **33** 249–254.

Emlen, J. T. (1971) Population densities of birds derived from transect counts. *Auk* **88** 323–341.

Errington, J. C. (1973) The effect of regular and random distributions on the analysis of pattern. *Journal of Ecology* **61** 99–105.

Fairburn, K. J. and Robinson, G. (1969) An application of trend surface mapping to the distribution of residuals from a regression. *Annals of the Association of American Geographers* **59** 158–170.

Fairfield Smith, H. (1938) An empirical law describing heterogeneity in the yields of agricultural crops. *Journal of Agricultural Science* **28** 1–23.

Fasham, M. J. R. (1978a) The statistical and mathematical analysis of plankton patchiness. *Oceanographic Marine Biology Annual Review* **16** 43–79.

Fasham, M. J. R. (1978b) The application of some stochastic processes to the study of plankton patchiness. *Spatial Pattern in Plankton Communities*. J. H. Steele, Ed. Plenum, New York. 131–156.

Finney, D. J. (1948) Random and systematic sampling in timber surveys. *Forestry* **22** 64–99.

Finney, D. J. (1950) An example of periodic variation in forest sampling. *Forestry* **23** 96–111.

Finney, D. J. (1953) The estimation of error in systematic sampling of forests. *Journal of Indian Society Agricultural Statistics* **5** 6–16.

Finney, D. J. and Pala, H. (1949) The elimination of bias due to edge effects in forest sampling. *Forestry* **23** 31–37.

Fisher, L. (1972) A survey of the mathematical theory of multidimensional point processes. *Stochastic Point Processes*. P. A. W. Lewis, Ed. Wiley, New York. 468–513.

Fisher, R. A., Thornton, H. G., and Mackenzie, W. A. (1922) The accuracy of the plating method of estimating the density of bacterial populations, with particular reference to the use of Thornton's agar medium with soil samples. *Annals of Applied Biology* **9** 325–359.

Fisher, W. D. (1971) Econometric estimation with spatial dependence. *Regional and Urban Economics* **1** 19–40.

Ford, E. D. (1976) The canopy of a Scots Pine forest: Description of a surface of complex roughness. *Agricultural Meteorology* **17** 9–32.

Fraser, A. R. (1977) Triangle based probability polygons for forest sampling. *Forestry Science* **23** 111–121.

Fraser, A. R. and van den Driessche, P. (1972) Triangles, density, and pattern in point populations. *IUFRO Third Conference Advisory Group of Forestry Statisticians*. Institut National de la Recherche Agronomique, Paris. 277–286.

Gates, C. E. (1969) Simulation study of estimators for the line transect sampling method. *Biometrics* **25** 317–328.

Gates, C. E. (1980) Line transect and related issues. *Sampling Biological Populations*. R. M. Cormack, G. P. Patil, and D. S. Robson, Eds. International Co-operative Publishing House, Burtonsville, Md. 71–154.

Gates, C. E., Marshall, W. H., and Olson, D. P. (1968) Line transect method of estimating grouse population densities. *Biometrics* **24** 135–145.

Geary, R. C. (1954) The contiguity ratio and statistical mapping. *The Incorporated Statistician* **5** 115–145.

Geciauskas, E. (1977) The distribution function of the distance between two points in a convex domain. *Advances in Applied Probability* **9** 427–428.

Getis, A. and Boots, B. (1978) *Models of Spatial Processes*. Cambridge University Press, London.

Giger, H. (1967) Ermittlung der mittleren Masszahlen von Partikeln eines Körpersystems durch Messungen auf dem Rand eines Schnittbereichs. *Zeitschrift für Angewandte Mathematik und Physik* **18** 883–888.

Gilbert, E. N. (1961) Random plane networks. *Journal of the Society for Industrial and Applied Mathematics* **9** 533–543.

Gilbert, E. N. (1962) Random subdivisions of space into crystals. *Annals of Mathematical Statistics* **33** 958–972.

Glass, L. and Tobler, W. R. (1971) Uniform distribution of objects in a homogeneous field: Cities on a plain. *Nature (London)* **233**(5314) 67–68.

Gleeson, A. C. and Douglas, J. B. (1975) Quadrat sampling and the estimation of Neyman type A and Thomas distributional parameters. *Australian Journal of Statistics* **17** 103–113.

Goldsmith, P. L. (1967) The calculation of true particle size distributions from the sizes observed in a thin slice. *British Journal of Applied Physics* **18** 813–830.

Goodall, D. W. (1963) Pattern analysis and minimal area—Some further comments. *Journal of Ecology* **51** 705–710.

Goodall, D. W. (1974) A new method for the analysis of spatial pattern by random pairing of quadrats. *Vegetatio* **29** 135–146.

Granger, C. W. J. (1969) Spatial data and time series analysis. *London Papers in Regional Science* **1** 1–24.

Grant, F. (1957) A problem in the analysis of geophysical data. *Geophysics* **22** 309–344.

Gray, J. M. (1972) Trend surface analysis: Trends through clusters. *Area* **4** 275–279.

Greco, A., Jeulin, D., and Serra, J. (1979) The use of the texture analyser to study sinter structure: Application to the morphology of calcium ferrites encountered in basic sinters of rich iron ores. *Journal of Microscopy* **116** 199–211.

Greeley, D. A. and Crapo, J. D. (1978) Practical approach to the estimation of the overall mean caliper diameter of a population of spheres and its application to data where small profiles are missed. *Journal of Microscopy* **114** 261–269.

Green, P. J. (1978) Small distances, and Monte Carlo testing of spatial pattern. *Advances in Applied Probability* **10** 493.

Green, P. J. and Sibson, R. (1978) Computing Dirichlet tessellations in the plane. *Computer Journal* **21** 168–173.

Green, R. H. (1966) Measurement of non-randomness in spatial distributions. *Researches on Population Ecology* **8** 1–7.

Gregory, G. G. (1977) Large sample theory for U-statistics and tests of fit. *Annals of Statistics* **5** 110–123.

Greig-Smith, P. (1952) The use of random and contiguous quadrats in the study of the structure of plant communities. *Annals of Botany* **16** 293–316.

Greig-Smith, P. (1961) Data on pattern within plant communities. I. The analysis of pattern. *Journal of Ecology* **49** 695–702.

Greig-Smith, P. (1964) *Quantitative Plant Ecology*. 2nd ed. Butterworths, London.

Greig-Smith, P. (1979) Pattern in vegetation. *Journal of Ecology* **67** 755–779.

Greig-Smith, P. and Chadwick, M. J. (1965) Data on pattern within plant communities. III. *Acacia-Capparis* semi-desert scrub in the Sudan. *Journal of Ecology* **53** 465–474.

Grenander, U. (1973) Statistical geometry: A tool for pattern analysis. *Bulletin of the American Mathematical Society* **79** 829–856.

Grenander, U. (1976, 1978) *Lectures in Pattern Theory. I. Pattern Synthesis. II. Pattern Analysis.* Springer-Verlag, New York.

Grosenbaugh, L. R. (1952a) Shortcuts for cruisers and scalers. *Occasional Papers, South Forestry Experimental Station* **126** 1–24.

Grosenbaugh, L. R. (1952b) Plotless timber estimates—New, fast, easy. *Journal of Forestry* **50** 32–37.

Guarascio, M., Huijbregts, C. J. and David, M. (Eds.) (1976) *Advanced Geostatistics in the Mining Industry.* Reidel, Dordrecht.

Gudgin, G. and Thomas, J. B. (1974) Probability in geographical research: Applications and problems. *The Statistician* **23** 157–177.

Guild, F. J. and Silverman, B. W. (1978) The microstructure of glass fibre reinforced polyester resin composites. *Journal of Microscopy* **114** 131–141.

Gulmon, S. L. and Mooney, H. A. (1977) Spatial and temporal relationships between two desert shrubs, *Atriplex hymenelytra* and *Tidestromia oblongifolia* in Death Valley, California. *Journal of Ecology* **65** 831–838.

Gundersen, H. J. G. (1977) Notes on the estimation of the numerical density of arbitrary profiles: The edge effect. *Journal of Microscopy* **111** 219–223.

Gundersen, H. J. G. (1978) Estimators of the number of objects per area unbiased by edge effects. *Microscopica Acta* **81** 107–117.

Gundersen, H. J. G., Jensen, T. B., and Østerby, R. (1978) Distribution of membrane thickness determined by lineal analysis. *Journal of Microscopy* **113** 27–43.

Haas, A., Matheron, G., and Serra, J. (1967) Morphologie mathématique et granulometries en place. *Annales des Mines* **XI** 736–753; **XII** 767–782.

Hadwiger, H. (1957) *Vorlesungen über Inhalt, Oberfläche und Isoperimetrie.* Springer-Verlag, Berlin.

Hafner, R. (1972) The asymptotic distribution of random clumps. *Computing* (Vienna) **10** 335–351.

Haggett, P., Cliff, A. D., and Frey, A. (1977) *Locational Analysis in Human Geography.* Arnold, London.

Haining, R. P. (1977) Model specification in stationary random fields. *Geographical Analysis* **9** 107–129.

Haining, R. P. (1978) The moving average model for spatial interaction. *Transactions, Institute of British Geographers N. S.* **3** 205–225.

Hall, J. B. (1971) Pattern in a chalk grassland community. *Journal of Ecology* **59** 749–762.

Hamilton, W. D. (1971) Geometry for the selfish herd. *Journal of Theoretical Biology* **31** 295–311.

Hammersley, J. M. and Handscomb, D. C. (1964) *Monte Carlo Methods.* Methuen, London.

Hannan, E. J. (1962) Systematic sampling. *Biometrika* **49** 281–283.

Harbaugh, J. W. and Preston, F. W. (1966) Fourier series analysis in geology. Reprinted in *Spatial Analysis.* B. J. L. Berry and D. F. Marble, Eds. Prentice-Hall, Englewood Cliffs, N.J. 218–238.

Harvey, D. W. (1966) Geographical processes and the analysis of point patterns: Testing models of diffusion by quadrat sampling. *Transactions and Papers, Institute of British Geographers* **40** 81–95.

Hasel, A. A. (1938) Sampling error in timber surveys. *Journal of Agricultural Research* **57** 713–736.

Hastings, W. K. (1970) Monte Carlo sampling methods using Markov chains and their applications. *Biometrika* **57** 97–109.

Hayne, D. W. (1949) An examination of the strip census method for estimating animal populations. *Journal of Wildlife Management* **13** 145–157.

Heine, V. (1955) Models for two-dimensional stationary stochastic processes. *Biometrika* **42** 170–178.

Hepple, L. W. (1974) The impact of stochastic process theory upon spatial analysis in human geography. *Progress in Geography* **6** 89–142.

Hepple, L. W. (1976) A maximum likelihood model for econometric estimation with spatial series. *London Papers in Regional Science* **6** 90–104.

Hepple, L. W. (1979) Bayesian analysis of the linear model with spatial dependence. *Exploratory and Explanatory Statistical Analysis of Spatial Data.* C. P. A. Bartels and R. H. Ketellapper, Eds. Martinus Nijhoff, Boston, Mass. 179–199.

Hessing, R. C., Lee, H. K., Pierce, A., and Powers, E. N. (1972) Automatic contouring using bicubic functions. *Geophysics* **37** 669–674.

Hill, M. O. (1973) The intensity of spatial pattern in plant communities. *Journal of Ecology* **61** 225–235.

Hilliard, J. E. (1972) Stereology: An experimental viewpoint. *Supplement Advances in Applied Probability* **4** 92–111.

Hines, W. G. S., and Hines, R. J. O'Hara (1979) The Eberhardt statistic, and the detection of nonrandomness of spatial point distributions. *Biometrika* **66** 73–79.

Hodder, I. and Orton, C. (1976) *Spatial Analysis in Archaeology.* Cambridge University Press, London.

Holgate, P. (1964) The efficiency of nearest neighbour estimators. *Biometrics* **20** 647–649.

Holgate, P. (1965a) Some new tests of randomness. *Journal of Ecology* **53** 261–266.

Holgate, P. (1965b) Tests of randomness based on distance methods. *Biometrika* **52** 345–353.

Holgate, P. (1965c) The distance from a random point to the nearest point of a closely packed lattice. *Biometrika* **52** 261–263.

Holgate, P. (1967) The angle-count method. *Biometrika* **54** 615–623.

Holgate, P. (1972) The use of distance methods for the analysis of spatial distribution of points. *Stochastic Point Processes.* P. A. W. Lewis, Ed. Wiley, New York. 122–135.

Holmes, J. H. (1970) The theory of plane sampling and its application in geographic research. *Economic Geography* **46** 379–392.

Hope, A. C. A. (1968) A simplified Monte Carlo significance test procedure. *Journal of the Royal Statistical Society* **B30** 582–598.

Hopkins, B. (1954) A new method of determining the type of distribution of plant individuals. *Annals of Botany* **18** 213–227.

Hordijk, L. (1974) Spatial correlation in the disturbances of a linear interregional model. *Regional and Urban Economics* **4** 117–140.

Hordijk, L. and Nijkamp, P. (1977) Dynamic models of spatial autocorrelation. *Environment and Planning* **A9** 505–520.

Howarth, R. J. (1967) Trend-surface fitting to random data–An experimental test. *American Journal of Science* **265** 619–625.

Howell, P. R., Fleet, D. E., Hildon, A., and Ralph, B. (1976) Field-ion microscopy of segregation to planar imperfections. *Journal of Microscopy* **107** 155–167.

Hsu, S. Y. and Mason, J. D. (1974) The nearest neighbor statistics for testing randomness of point distributions in a bounded two-dimensional space. *Proceedings of the 1972 Meeting of the IGU Commission on Quantitative Geography.* M. H. Yeates, Ed. McGill–Queen's University Press, Montreal.

Huijbregts, C. J. (1975) Regionalized variables and quantitative analysis of spatial data. *Display and Analysis of Spatial Data.* J. C. Davis and M. J. McCullagh, Eds. Wiley, New York. 38–53.

Iwao, S. (1972) Application of the $\dot{m}-m$ method to the analysis of spatial patterns by changing the quadrat size. *Researches on Population Ecology* **14** 97–128.

Iwao, S. and Kuno, E. (1971) An approach to the analysis of aggregation pattern in biological populations. *Statistical Ecology.* G. P. Patil, E. C. Pielou, and W. E. Waters, Eds. Pennsylvania State University Press, College Town, Pa. I 461–513.

Jakeman, A. J. and Anderssen, R. S. (1975) Abel type integral equations in stereology. I. General discussion. *Journal of Microscopy* **105** 121–133.

Jensen, E. B., Gundersen, H. J. G., and Østerby, R. (1979) Determination of membrane thickness distribution from orthogonal intercepts. *Journal of Microscopy* **115** 19–33.

Johnson, G. G. and Vance, V. (1967) Application of a Fourier data-smoothing technique to the meteoric crater Ries Kessell. *Journal of Geographical Research* **72** 1741–1750.

Jolivet, E. (1978) Caractérisation et test du caractère agrégatif des processus ponctuels stationnaires sur R^2. *Lecture Notes in Mathematics* **636** 1–25.

Jolivet, E. (1980) Théorème centrale limite et convergence de processus empiriques pour des processus ponctuels stationnaires. *Point Processes and Queueing Systems.* P. Bártfai and J. Tomkó, Eds. North Holland, Amsterdam.

Journel, A. G. (1977) Kriging in terms of projections. *Mathematical Geology* **9** 563–586.

Journel, A. G. and Huijbregts, C. J. (1978) *Mining Geostatistics.* Academic Press, London.

Jowett, G. H. (1955) Least squares regression analysis for trend-reduced time series. *Journal of the Royal Statistical Society* **B17** 91–104.

Julesz, B. (1975) Experiments in the visual perception of texture. *Scientific American* **April** 34–43.

Jumars, P. A. (1978) Spatial autocorrelation with RUM: Vertical and horizontal structure of a bathyal benthic community. *Deep-Sea Research* **25** 589–604.

Jumars, P. A., Thistle, D., and Jones, M. L. (1977) Detecting two-dimensional structure in biological data. *Oecologica* **28** 109–123.

Kallenberg, O. (1976) *Random Measures.* Akademie-Verlag, Berlin.

Keiding, N., Jensen, S. T., and Ranek, L. (1972) Maximum likelihood estimation of the size distribution of liver cell nuclei from the observed distribution in a plane section. *Biometrics* **28** 813–829.

Kelly, F. P. and Ripley, B. D. (1976) A note on Strauss's model for clustering. *Biometrika* **63** 357–360.

Kendall, D. G. (1974) Foundations of a theory of random sets. *Stochastic Geometry.* E. F. Harding and D. G. Kendall, Eds. Wiley, Chichester 322–376.

Kendall, M. G. and Moran, P. A. P. (1963) *Geometrical Probability.* Griffin, London.

Kennedy, J. S. and Crawley, L. (1967) Spaced-out gregariousness in sycamore aphids *Drepanosiphum platanoides* (Schank). *Journal of Animal Ecology* **36** 147–170.

Kershaw, K. A. (1957) The use of cover and frequency in the detection of pattern in plant communities. *Ecology* **38** 291–299.

Kershaw, K. A. (1958, 1959) An investigation of the structure of a grassland community. I. The pattern of *Agrostis tenuis*. II. The pattern of *Dactylis glomerata, Lolium perenne* and *Trifolium repens*. III. Discussion and conclusions. *Journal of Ecology* **46** 571–592; **47** 31–53.

Kershaw, K. A. (1960) The detection of pattern and association. *Journal of Ecology* **48** 233–242.

Kershaw, K. A. (1961) Association and co-variance analysis of plant communities. *Journal of Ecology* **49** 643–654.

Kershaw, K. A. (1973) *Quantitative and Dynamic Plant Ecology*. 2nd ed. Arnold, London.

Kester, A. (1975) Asymptotic normality of the number of small distances between random points in a cube. *Stochastic Processes and their Applications* **3** 45–54.

Keuls, M., Over, H. J., and de Wit, C. T. (1963) The distance method for estimating densities. *Statistica Neerlandica* **17** 71–91.

Kiang, T. (1966) Random fragmentation in two and three dimensions. *Zeitschrift für Astrophysik* **64** 433–439.

King, C. A. M. (1969) Trend surface analysis of Central Pennine erosion surfaces. *Institute of British Geographers Transactions* **47** 47–59.

King, L. J. (1962) A quantitative expression of the pattern of urban settlements in selected areas of the United States. *Tijdschrift voor Economische en Sociale Geografie* **53** 1–7. Reprinted in *Spatial Analysis*. B. J. L. Berry and D. F. Marble, Eds. Prentice-Hall, Englewood Cliffs, N. J. 159–164.

Klein, J. C. and Serra, J. (1972) The texture analyser. *Journal of Microscopy* **95** 349–356.

Kovner, J. L. and Patil, S. A. (1974) Properties of estimators of wildlife population density for the line transect method. *Biometrics* **30** 225–230.

Krige, D. G. (1966) Two-dimensional weighted moving average trend surfaces for ore evaluation. *Proceedings of the Symposium on Mathematical Statistics and Computer Applications in Ore Valuation, Johannesburg*. 13–38.

Krishna Iyer, P. V. (1949) The first and second moments of some probability distributions arising from points on a lattice and their applications. *Biometrika* **36** 135–141.

Krumbein, W. C. (1956) Regional and local components in facies maps. *Bulletin of the American Association of Petroleum Geologists* **40** 2163–2194.

Krumbein, W. C. (1959) Trend surface analysis of contour-type maps with irregular control-point spacing. *Journal of Geophysical Research* **64** 823–834.

Krumbein, W. C. (1963) Confidence intervals for low-order polynomial trend surfaces. *Journal of Geophysical Research* **68** 5869–5878.

Lantuejoul, C. (1978) Computation of the histograms of the number of edges and neighbours of cells in a tessellation. *Lecture Notes in Biomathematics* **23** 323–329.

Lauritzen, S. L. (1973) *The Probabilistic Background of Some Statistical Methods in Physical Geodesy*. Geodaetisk Institut Meddelelse København. **48** 1–96.

Lawrence, P. A. (1969) Cellular differentiation and pattern formation during metamorphosis of the milkweed bug *Oncopeltus*. *Developmental Biology* **19** 12–40.

Lawson, C. L. (1977) Software for C^1 surface interpolation. *Mathematical Software III* J. R. Rice, Ed. Academic Press, New York. 161–194.

Lebart, L. (1969) Analyse statistique de la contiguité. *Publications de l'Institut de Statistique de l'Université de Paris* **18(2)** 81–112.

Leese, J. A. and Epstein, E. S. (1963) Application of two-dimensional spectral analysis to the quantification of satellite cloud photographs. *Journal of Applied Meteorology* **2** 629–644.

Lehmann, E. L. (1975) *Nonparametrics. Statistical Methods Based on Ranks.* Holden-Day, San Francisco, Calif.

Lesseps, R. J., Geurts, A. H. M. van Kessel, and Denuce, J. M. (1975) Cell patterns and cell movements during early development of an annual fish, *Nothobranchius neumanni*. *Journal of Experimental Zoology* **193** 137–146.

Lewis, P. A. W. (Ed.) (1972) *Stochastic Point Processes.* Wiley, New York.

Lewis, S. M. (1975) Robust estimation of density for a two-dimensional point process. *Biometrika* **62** 519–521.

Liebetrau, A. M. (1977) Tests of randomness in two dimensions. *Communications in Statistics —Theory and Methods* **A6(14)** 1367–1383.

Liebetrau, A. M. (1978) The weak convergence of a class of estimators of the variance function of a two-dimensional Poisson process. *Journal of Applied Probability* **15** 433–439.

Liebetrau, A. M. and Rothman, E. D. (1977) A classification for spatial distributions based on several cell sizes. *Geographical Analysis* **9** 14–28.

Link, R. F. and Koch, G. S. (1970) Experimental designs and trend-surface analysis. *Geostatistics.* D. F. Merriam, Ed. Plenum, New York. 23–29.

Lloyd, M. (1967) Mean crowding. *Journal of Animal Ecology* **36** 1–30.

Longuet-Higgins, M. S. (1957) The statistical analysis of a random, moving surface. *Philosophical Transactions of the Royal Society of London* **A249** 321–387.

McConalogue, D. J. (1970) A quasi-intrinsic scheme for passing a smooth curve through a discrete set of points. *Computer Journal* **13** 392–396.

McConalogue, D. J. (1971) An automatic French-curve procedure for use with an incremental plotter. *Computer Journal* **14** 207–209.

McIntyre, G. A. (1953) Estimation of plant density using line transects. *Journal of Ecology* **41** 319–330.

McLain, D. H. (1974) Drawing contours from arbitrary data points. *Computer Journal* **17** 318–324.

McLain, D. H. (1976) Two dimensional interpolation from random data. *Computer Journal* **19** 178–181; 384.

Mandelbrot, B. B. (1975) Sur un modèle décomposable d'univers hiérachisé: Déduction des corrélations galactiques sur la sphère céleste. *Compte Rendu Academiè Science Paris* **280A** 1551–1554.

Mandelbrot, B. B. (1977) *Fractals: Form, Chance and Dimension.* W. H. Freeman, San Francisco, Calif.

Mardia, K. V., Edwards, R., and Puri, M. L. (1978) Analysis of central place theory. *Bulletin of the International Statistical Institute* **47(2)** 93–110.

Maréchal, A. (1975) Geostatistique et applications minières. *Annales des Mines* Nov. 27–38.

Maréchal, A. (1976) The practice of transfer functions: Numerical methods and their application. *Advanced Geostatistics in the Mining Industry.* M. Guarascio, C. J. Huijbregts, and M. David, Eds. Reidel, Dordrecht. 253–276.

Maréchal, A. and Serra, J. (1970) Random kriging. *Geostatistics.* D. F. Merriam, Ed. Plenum, New York. 91–112.

Mark, A. F. and Esler, A. E. (1970) An assessment of the point-centred quarter method of plotless sampling in some New Zealand forests. *Proceedings of the New Zealand Ecological Society* **17** 106–110.

Marlow, S. and Powell, M. J. D. (1976) A Fortran subroutine for plotting the part of a conic that is inside a given triangle. *UKAEA Harwell report* AERE-R8336.

Marquiss, M., Newton, I., and Radcliffe, D. A. (1978) The decline of the raven, *Corvus corax*, in relation to afforestation in Southern Scotland and Northern England. *Journal of Applied Ecology* **15** 129–144.

Marriott, F. H. C. (1972) Buffon's problems for non-random distribution. *Biometrics* **28** 621–624.

Marriott, F. H. C. (1979) Monte-Carlo tests: How many simulations? *Applied Statistics* **28** 75–77.

Martin, F. B. (1973) Beehive designs for observing variety competition. *Biometrics* **29** 397–402.

Martin, R. J. (1979) A subclass of lattice processes applied to a problem in planar sampling. *Biometrika* **66** 209–217.

Martin, R. L. (1974) On autocorrelation, bias and the use of first spatial difference in regression analysis. *Area* **6** 185–194.

Martin, R. L. and Oeppen, J. E. (1975) The identification of regional forecasting models using space–time correlation functions. *Transactions, Institute of British Geographers* **66** 95–118.

Matérn, B. (1960) *Spatial Variation*. Meddelanden från Statens Skogsforskningsinstitut **49,5.** 1–144.

Matérn, B. (1971) Doubly stochastic Poisson processes in the plane. *Statistical Ecology*. G. P. Patil, E. C. Pielou, and W. E. Waters, Eds. Pennsylvania State University Press, College Town, Pa. **I** 195–213.

Matérn, B. (1972) Poisson processes in the plane and related models for clumping and heterogeneity. *NATO Advanced Study Institute on Statistical Ecology, Pennsylvania State University*.

Matheron, G. (1963) Principles of geostatistics. *Economic Geology* **58** 1246–1266.

Matheron, G. (1965) *Les Variables Régionalisées et leur Estimation*. Masson, Paris.

Matheron, G. (1967a) *Eléments pour une Théorie des Milieux Poreux*. Masson, Paris.

Matheron, G. (1967b) Kriging, or polynomial interpolation procedures. *Canadian Mining and Metallurgical Bulletin* **60** 1041–1045.

Matheron, G. (1970) Structures aléatoires et géologie mathématiques. *International Statistical Review* **38** 1–11.

Matheron, G. (1972a) Ensembles fermés aléatoires, ensembles semi-markoviens et polyèdres poissoniens. *Advances in Applied Probability* **4** 508–541.

Matheron, G. (1972b) Random sets theory and its applications to stereology. *Journal of Microscopy* **95** 15–23, 393.

Matheron, G. (1973) The intrinsic random functions and their applications. *Advances in Applied Probability* **5** 439–468.

Matheron, G. (1975) *Random Sets and Integral Geometry*. Wiley, New York.

Matheron, G. (1976a) A simple substitute for conditional expectation: The disjunctive kriging. *Advanced Geostatistics in the Mining Industry*. M. Guarascio, C. J. Huijbregts, and M. David, Eds. Reidel, Dordrecht. 221–236.

Matheron, G. (1976b) Forecasting block grade distributions: The transfer functions. *Ad-*

vanced Geostatistics in the Mining Industry. M. Guarascio, C. J. Huijbregts, and M. David, Eds. Reidel, Dordrecht. 237–251.

Matthes, K., Kerstan, J., and Mecke, J. (1978) *Infinitely Divisible Point Processes*. Wiley, Chichester.

Maude, A. D. (1973) Interpolation—Mainly for graph plotters. *Computer Journal* **16** 64–65.

Mawson, J. C. (1968) A Monte Carlo study of distance measures in sampling for spatial distribution in forest stands. *Forestry Science* **14** 127–139.

Mayhew, T. M. and Cruz Orive, L.-M. (1974) Caveat on the use of the Delesse principle of areal analysis for estimating component volume densities. *Journal of Microscopy* **102** 195–207.

Mead, R. (1966) A relationship between individual plant-spacing and yield. *Annals of Botany* **30** 301–309.

Mead, R. (1967) A mathematical model for the estimation of inter-plant competition. *Biometrics* **23** 189–205.

Mead, R. (1968) Measurement of competition between individual plants in a population. *Journal of Ecology* **56** 35–45.

Mead, R. (1971) Models for interplant competition in irregularly spaced populations. *Statistical Ecology*. G. P. Patil, E. C. Pielou and W. E. Waters, Eds. Pennsylvania State University Press, College Town, Pa. **II** 13–32.

Mead, R. (1974) A test for spatial pattern at several scales using data from a grid of contiguous quadrats. *Biometrics* **30** 295–307.

Mead, R. (1978) Letter to the editor. *Biometrics* **34** 714–716.

Medvedkov, Y. V. (1967) The concept of entropy in settlement pattern analysis. *Regional Science Association Papers* **18** 165–168.

Mercer, W. B. and Hall, A. D. (1911) The experimental error of field trials. *Journal of Agricultural Science* **4** 107–132.

Miesch, A. T. (1975) Variograms and variance components in geochemistry and ore evaluation. *Geological Society of America Memoirs* **142** 333–340.

Miles, R. E. (1964) Random polygons determined by lines in a plane. I and II. *Proceedings of the National Academy of Sciences* **52** 901–907; 1157–1160.

Miles, R. E. (1969) Poisson flats in Euclidean spaces. Part I. A finite number of random uniform flats. *Advances in Applied Probability* **1** 211–237.

Miles, R. E. (1970) On the homogeneous planar Poisson point process. *Mathematical Biosciences* **6** 85–127.

Miles, R. E. (1971) Poisson flats in Euclidean spaces. Part II. Homogeneous Poisson flats and the complementary theorem. *Advances in Applied Probability* **3** 1–43.

Miles, R. E. (1972a) Multidimensional perspectives on stereology. *Journal of Microscopy* **95** 181–195.

Miles, R. E. (1972b) The random division of space. *Supplement Advances in Applied Probability* **4** 243–266.

Miles, R. E. (1973) The various aggregates of random polygons determined by random lines in a plane. *Advances in Mathematics*, **10** 256–290.

Miles, R. E. (1974a) A synopsis of "Poisson flats in Euclidean spaces." *Stochastic Geometry*. E. F. Harding and D. G. Kendall, Eds. Wiley, Chichester. 202–227.

Miles, R. E. (1974b) On the elimination of edge effects in planar sampling. *Stochastic Geometry*. E. F. Harding and D. G. Kendall, Eds. Wiley, Chichester. 228–247.

Miles, R. E. (1975) Direct derivations of certain surface integral formulae for the mean projections of a convex set. *Advances in Applied Probability* **7** 818–829.

Miles, R. E. (1978a) The sampling, by quadrats, of planar aggregates. *Journal of Microscopy* **113** 257–267.

Miles, R. E. (1978b) The importance of proper model specification in stereology. *Lecture Notes in Biomathematics* **23** 115–136.

Miles, R. E. and Davy, P. (1976) Precise and general conditions for the validity of a comprehensive set of stereological fundamental formulae. *Journal of Microscopy* **107** 211–226.

Miles, R. E. and Davy, P. (1977) On the choice of quadrats in stereology. *Journal of Microscopy* **110** 27–44.

Miles, R. E. and Davy, P. (1978) Particle number or density can be stereologically estimated by wedge sections. *Journal of Microscopy* **113** 45–52.

Miles, R. E. and Fischer, R. A. (1973) The role of spatial pattern in the competition between crop plants and weeds. A theoretical analysis. *Mathematical Biosciences* **18** 335–350.

Miles, R. E. and Serra, J. (Eds.) (1978) *Geometrical Probability and Biological Structures: Buffon's 200th Anniversary*. Lecture Notes in Biomathematics **23**. 1–338.

Millier, C., Poissonnet, M., and Serra, J. (1972) Morphologie mathématique et sylviculture. *IUFRO Third Conference Advisory Group Forest Statisticians*. Institut National de la Recherche Agnomomique, Paris. 287–307.

Milne, A. (1959) The centric systematic area-sample treated as a random sample. *Biometrics* **15** 270–297.

Moellering, H. and Tobler, W. R. (1972) Geographical variances. *Geographical Analysis* **4** 34–50.

Mollison, D. (1977) Spatial contact models for ecological and epidemic spread. *Journal of the Royal Statistical Society* **B39** 283–326.

Moore, P. G. (1953) A test for non-randomness in plant populations. *Annals of Botany* **17** 57–62.

Moore, P. G. (1954) Spacing in plant populations. *Ecology* **35** 222–227.

Moore, P. (1955) The strip intersect census. *Transactions of the North American Wildlife Conference* **20** 390–405.

Moran, P. A. P. (1948) The interpretation of statistical maps. *Journal of the Royal Statistical Society* **B10** 243–251.

Moran, P. A. P. (1950) Notes on continuous stochastic phenomena. *Biometrika* **37** 17–23.

Moran, P. A. P. (1972) The probabilistic basis of stereology. *Supplement Advances in Applied Probability* **4** 69–91.

Moran, P. A. P. (1973a) A Gaussian Markovian process on a square lattice. *Journal of Applied Probability* **10** 54–62.

Moran, P. A. P. (1973b) Necessary conditions for Markovian processes on a lattice. *Journal of Applied Probability* **10** 605–612.

Morisita, M. (1954) Estimation of population density by a spacing method. *Memoirs of the Faculty of Science, Kyushu University (E)* **1** 187–197.

Morisita, M. (1957) A new method for the estimation of density by the spacing method applicable to non-randomly distributed populations. *Physiological Ecology Kyoto* **7** 134–144.

Morisita, M. (1959) Measuring the dispersion of individuals and analysis of distribution patterns. *Memoirs of the Faculty of Science, Kyushu University* (*E*) **2** 215–235.

Morisita, M. (1962) I_δ index, a measure of dispersion of individuals. *Researches on Population Ecology* **4** 1–7.

Morisita, M. (1964) Application of I_δ index to sampling techniques. *Researches on Population Ecology* **6** 43–53.

Morisita, M. (1971) Composition of the I_δ index. *Researches on Population Ecology* **13** 1–27.

Morton, A. J. (1974) Ecological studies of a fixed dune grassland at Newborough Warren, Anglesey. I. The structure of the grassland. II. Causal factors of grassland structure. *Journal of Ecology* **62** 253–278.

Mountford, M. D. (1961) On E. C. Pielou's index of non-randomness. *Journal of Ecology* **49** 271–275.

Moussouris, J. (1974) Gibbs and Markov systems with constraints. *Journal of Statistical Physics* **10** 11–33.

Nelder, J. A. and Wedderburn, R. W. M. (1972) Generalized linear models. *Journal of the Royal Statistical Society* **A135** 370–384.

Neveu, J. (1965) *Mathematical Foundations of the Calculus of Probability*. Holden-Day, San Francisco, Ca.

Newton, I., Marquiss, M., Weir, D. N., and Moss, D. (1977) Spacing of sparrowhawk nesting territories. *Journal of Animal Ecology* **46** 425–441.

Neyman, J. and Scott, E. L. (1972) Processes of clustering and applications. *Stochastic Point Processes*. P. A. W. Lewis, Ed. Wiley, New York. 646–681.

Nicholson, W. L. (1970) Estimation of linear properties of particle size distributions. *Biometrika* **57** 273–297.

Nicholson, W. L. (Ed.) (1972) *Proceedings of the Symposium on Statistical and Probabilistic Problems in Metallurgy, Seattle, 1971. Supplement to Advances in Applied Probability* **4** 1–266.

Nicholson, W. L. (1976) Estimation of linear functionals by maximum likelihood. *Journal of Microscopy* **107** 323–334.

Nicholson, W. L. and Merckx, K. R. (1969) Unfolding particle size distributions. *Technometrics* **11** 707–723.

Noether, G. E. (1970) A central limit theorem with nonparametric applications. *Annals of Mathematical Statistics* **41** 1753–1755.

Norcliffe, G. B. (1969) On the use and limitations of trend surface models. *Canadian Geographer* **13** 338–348.

Olea, R. A. (1974) Optimal contour mapping using universal kriging. *Journal of Geophysical Research* **79** 695–702; **80** 832–836.

Olsson, G. (1968) Complementary models: A study of colonisation maps. *Geografiska Annaler* **50B** 115–132.

Ord, K. (1975) Estimation methods for models of spatial interaction. *Journal of the American Statistical Association* **70** 120–126.

Ord, K. (1978) How many trees in a forest? *Mathematical Scientist* **3** 23–33.

Orloci, L. (1971) An information theory model for pattern analysis. *Journal of Ecology* **59** 343–349.

Osbourne, J. G. (1942) Sampling errors of systematic and random surveys of cover-type areas. *Journal of the American Statistical Association* **37** 256–264.

Owen, M. and Harberd, D. J. (1970) Vegetational pattern in a stable grassland community. *Journal of Ecology* **58** 399–408.

Paloheimo, J. E. (1971) On a theory of search. *Biometrika* **58** 61–75.

Paloheimo, J. E. and Vukov, A. M. (1976) On measures of aggregation and indices of contagion. *Mathematical Biosciences* **30** 69–97.

Papadakis, J. S. (1937) Méthode statistique pour des expériences sur champ. *Bulletin de l'Institut Amél. Plantes à Salonique* **23**.

Parker, K. R. (1979) Density estimation by variable area transect. *Journal of Wildlife Management* **43** 484–492.

Parsley, A. J. (1971) Application of autocorrelation criteria to the analysis of mapped geologic data from the coal measures of Central England. *Mathematical Geology* **3** 281–295.

Patankar, V. N. (1954) The goodness of fit of frequency distributions obtained from stochastic processes. *Biometrika* **41** 450–462.

Patil, G. P. and Stiteler, W. M. (1974) Concepts of aggregation and their quantification: A critical review with some new results and applications. *Researches on Population Ecology* **15** 238–254.

Patil, G. P., Pielou, E. C., and Waters, W. E. (Eds.) (1971) *Statistical Ecology.* Proceedings of the International Symposium on Statistical Ecology, New Haven, 1969. Pennsylvania State University Press, College Town, Pa.

Payandeh, B. (1970a) Relative efficiency of two-dimensional systematic sampling. *Forestry Science* **16** 271–276.

Payandeh, B. (1970b) Comparison of methods for assessing spatial distributions of trees. *Forestry Science* **16** 312–317.

Peebles, P. J. E. (1973) Statistical analysis of catalogs of extragalactic objects. I. Theory. *Astrophysical Journal* **185** 413–440.

Peebles, P. J. E. (1974) The nature of the distribution of galaxies. *Astronomy and Astrophysics* **32** 197–202.

Peebles, P. J. E. and Groth, E. J. (1975) Statistical analysis of extragalactic objects. V. Three-point correlation function for the galaxy distribution in the Zwicky catalog. *Astrophysical Journal* **196** 1–11.

Pelto, C. R., Elkins, T. A., and Boyd, H. A. (1968) Automatic contouring of irregularly spaced data. *Geophysics* **33** 424–430.

Pemadasa, M. A., Greig-Smith, P., and Lovell, P. H. (1974) A quantitative description of the distribution of annuals in the dune system at Aberffaw, Anglesey. *Journal of Ecology* **62** 379–402.

Perry, J. N. and Mead, R. (1979) On the power of the dispersion test to detect spatial pattern. *Biometrics* **35** 613–622.

Persson, O. (1964) Distance methods. *Studia Forestalia Suecica* **15** 1–68.

Persson, O. (1965) Distance methods II. *Forest Research Institute of Sweden, Dept. of Forest Biometry Research Notes* **6** 1–24.

Persson, O. (1971) The robustness of estimating density by distance measurements. *Statistical Ecology.* G. P. Patil, E. C. Pielou, and W. E. Waters, Eds. Pennsylvania State University Press, College Town, Pa. **II** 175–190.

Persson, O. (1972) The border effect on the distance between sample point and closest individual in a square. *IUFRO Third Conference Advisory Group of Forest Statisticians.* Institut National de la Recherche Agronomique, Paris. 241–246.

Peskun, P. H. (1973) Optimal Monte-Carlo sampling using Markov chains. *Biometrika* **60** 607–612.

Pielou, E. C. (1957) The effect of quadrat size on the estimation of the parameters of Neyman's and Thomas's distributions. *Journal of Ecology* **45** 31–47.

Pielou, E. C. (1959) The use of point-to-plant distances in the study of the pattern of plant populations. *Journal of Ecology* **47** 607–613.

Pielou, E. C. (1960) A single mechanism to account for regular, random and aggregated populations. *Journal of Ecology* **48** 575–584.

Pielou, E. C. (1962) The use of plant-to-neighbour distances for the detection of competition. *Journal of Ecology* **50** 357–367.

Pielou, E. C. (1964) The spatial pattern of two-phase patchworks of vegetation. *Biometrics* **20** 156–167; 891–892.

Pielou, E. C. (1965) The concept of randomness in patterns of mosaics. *Biometrics* **21** 908–920.

Pielou, E. C. (1974) *Population and Community Ecology—Principles and Methods*. Gordon and Breach, New York.

Pielou, E. C. (1977) *Mathematical Ecology*. Wiley, New York.

Pike, M. C. and Smith, P. G. (1968) Disease clustering: A generalization of Knox's approach to the detection of space-time interactions. *Biometrics* **24** 541–556.

Pinder, D. A. and Witherick, M. E. (1972) The principles, practice and pitfalls of nearest neighbour analysis. *Geography* **57** 277–288.

Pollard, J. H. (1971) On distance estimators of density in randomly distributed forests. *Biometrics* **27** 991–1002.

Pollock, K. H. (1978) A family of density estimators for line transect sampling. *Biometrics* **34** 475–478.

Powell, M. J. D. and Sabin, M. A. (1977) Piecewise quadratic approximations on triangles. *ACM Transactions on Mathematical Software* **3** 316–325.

Prentice, R. L. (1973) A design for studying the clustering of plant or animal species using quadrat sizes in geometric progression. *Journal of Theoretical Biology* **39** 601–608.

Preston, C. J. (1974) *Gibbs States on Countable Sets*. Cambridge University Press, London.

Preston, C. J. (1976) *Random Fields*. Lecture Notes in Mathematics **534** 1–200.

Preston, C. J. (1977) Spatial birth-and-death processes. *Bulletin of the International Statistical Institute* **46(2)** 371–391.

Quenouille, M. H. (1949) Problems in plane sampling. *Annals of Mathematical Statistics* **20** 355–375.

Ramsey, F. L. (1979) Parametric models for line transect surveys. *Biometrika* **66** 505–512.

Rayner, J. N. (1971) *An Introduction to Spectral Analysis*. Pion, London.

Rayner, J. N. and Golledge, R. G. (1972) Spectral analysis of settlement patterns in diverse physical and economic environments. *Environment and Planning* **4** 347–371.

Rayner, J. N. and Golledge, R. G. (1973) The spectrum of US Route 40 re-examined. *Geographical Analysis* **5** 338–350.

Reilly, W. I. (1969) Contouring gravity anomaly maps by digital plotter. *New Zealand Journal of Geology and Geophysics* **12** 628–632.

Rhind, D. (1975) A skeletal overview of spatial interpolation techniques. *Computer Applications* **2** 293–309.

Rhodda, J. C. (1970) A trend surface analysis trial for the plantation surfaces of North Cardiganshire. *Institute of British Geographers Transactions* **50** 107–114.

Riechert, S. E. (1974) The pattern of local web distributions in a desert spider: Mechanisms and seasonal variation. *Journal of Animal Ecology* **43** 733–746.

Riechert, S. E., Reeder, W. G., and Allen, T. A. (1973) Patterns of spider distribution [*Agelenopsis aperta* (Gertsch)] in desert grassland and recent lava bed habitats, South-Central New Mexico. *Journal of Animal Ecology* **42** 19–35.

Ripley, B. D. (1976a) The second-order analysis of stationary point processes. *Journal of Applied Probability* **13** 255–266.

Ripley, B. D. (1976b) Locally finite random sets; Foundations for point process theory. *Annals of Probability* **4** 983–994.

Ripley, B. D. (1977) Modelling spatial patterns. *Journal of the Royal Statistical Society* **B39** 172–212.

Ripley, B. D. (1978) Spectral analysis and the analysis of pattern in plant communities. *Journal of Ecology* **66** 965–981.

Ripley, B. D. (1979a) Simulating spatial patterns: Dependent samples from a multivariate density. *Applied Statistics* **28** 109–112.

Ripley, B. D. (1979b) Tests of "randomness" for spatial point patterns. *Journal of the Royal Statistical Society* **B41** 368–374.

Ripley, B. D. (1979c) The analysis of geographical maps. *Exploratory and Explanatory Statistical Analysis of Spatial Data*. C. P. A. Bartels and R. H. Ketellapper, Eds. Martinus Nijhoff, Boston, Mass. 53–72.

Ripley, B. D. and Kelly, F. P. (1977) Markov point processes. *Journal of the London Mathematical Society* **15** 188–192.

Ripley, B. D. and Rasson, J.-P. (1977) Finding the edge of a Poisson forest. *Journal of Applied Probability* **14** 483–491.

Ripley, B. D. and Silverman, B. W. (1978) Quick tests for spatial regularity. *Biometrika* **65** 641–642.

Robinette, W. L., Jones, D. A., Gashwiler, G. S., and Aldous, C. M. (1954) Methods for censusing winter-lost deer. *Transactions of the North American Wildlife Conference* **19** 511–525.

Robinette, W. L., Jones, D. A., Gashwiler, G. S., and Aldous, C. M. (1956) Further analysis of methods of censusing winter-lost deer. *Journal of Wildlife Management* **20** 75–78.

Robinette, W. L., Loveless, C. M., and Jones, D. A. (1974) Field tests of strip census methods. *Journal of Wildlife Management* **38** 81–96.

Robinson, G. (1970) Some comments on trend surface analysis *Area* **2** 31–36.

Robinson, G. (1972) Trials on trends through clusters of cirques. *Area* **4** 102–103.

Robinson, J. E. (1970) Spatial filtering of geologic data. *International Statistical Review* **38** 21–34.

Roder, W. (1975) A procedure for assessing point patterns without reference to area or density. *The Professional Geographer* **27** 432–440.

Rogers, A. (1974) *Statistical Analysis of Spatial Dispersion. The Quadrat Method*. Pion, London.

Rogers, C. A. (1964) *Packing and Covering*. Cambridge University Press, London.

Rozanov, Yu. A. (1967) On the Gaussian homogeneous fields with given conditional distribution. *Theory of Probability and its Applications* **12** 381–391.

Ruelle, D. (1969) *Statistical Mechanics.* W. A. Benjamin, New York.

Ruelle, D. (1970) Superstable interaction in classical statistical mechanics. *Communications in Mathematical Physics* **18** 127–159.

Saltykov, S. A. (1967) The determination of the size distribution of particles in an opaque material from a measurement of the size distribution of their sections. *Stereology.* H. Elias, Ed. Springer, Berlin. 163–173.

Sands, W. A. (1965) Termite distribution in man-modified habitats in West Africa. *Journal of Animal Ecology* **34** 557–571.

Santaló, L. A. (1976) *Integral Geometry and Geometrical Probability.* Addison-Wesley, Reading, Mass.

Saunders, R. and Funk, G. M. (1977) Poisson limits for a clustering model of Strauss. *Journal of Applied Probability* **14** 776–784.

Schoenberg, I. J. (1938) Metric spaces and completely monotone functions. *Annals of Mathematics* **39** 811–841.

Schumaker, L. L. (1976) Fitting surfaces to scattered data. *Approximation Theory II.* G. G. Lorentz, C. K. Chui, and L. L. Schumaker, Eds., Academic Press, New York. 203–268.

Seber, G. A. F. (1976) *Linear Regression Analysis.* Wiley, New York.

Sen, A. K. (1976) Large sample-size distribution of statistics used in testing for spatial correlation. *Geographical Analysis* **9** 175–184.

Sen, A. R., Tourigny, J., and Smith, G. E. J. (1974) On the line transect sampling method. *Biometrics* **30** 329–340.

Sen, A. R., Smith, G. E. J., and Butler, G. (1978) On a basic assumption in the line transect method. *Biometrische Zeitschrift* **20** 363–369.

Serra, J. (1972) Stereology and structuring elements. *Journal of Microscopy* **95** 93–103.

Sexton, O. J. and Stalker, H. D. (1961) Spacing patterns of female *Drosophila paramelanica. Animal Behaviour* **9** 77–81.

Shanks, R. E. (1954) Plotless sampling trials in Appalachian forest types. *Ecology* **35** 237–244.

Shante, V. K. S. and Kilpatrick, S. (1971) An introduction to percolation theory. *Advances in Physics* **20** 325–357.

Shepard, D. (1968) A two-dimensional interpolation function for irregularly spaced data. *Proceedings 1968 ACM National Conference* 517–524.

Shepp, L. A. and Kruskal, J. B. (1978) Computerized tomography: The new medical X-ray technology. *American Mathematical Monthly* **85** 420–439.

Sibson, R. (1978) Locally equiangular triangulations. *Computer Journal* **21** 243–245.

Sibson, R. (1980a) The Dirichlet tessellation as an aid in data analysis. *Scandinavian Journal of Statistics* **7** 14–20.

Sibson, R. (1980b) A vector identity for the Dirichlet tessellation. *Mathematical Proceedings of the Cambridge Philosophical Society* **87** 151–155.

Sibson, R. (1980c) Natural neighbourhood interpolation, *Graphical Methods for Multivariate Data.* V. D. Barnett, Ed. Wiley, Chichester.

Silverman, B. W. (1976) Limit theorems for dissociated random variables. *Advances in Applied Probability* **8** 806–819.

Silverman, B. W. (1978) Distances on circles, toruses and spheres. *Journal of Applied Probability* **15** 136–143.

Silverman, B. and Brown, T. (1978) Short distances, flat triangles and Poisson limits. *Journal of Applied Probability* **15** 815–825.

Skellam, J. G. (1952) Studies in statistical ecology. I. Spatial pattern, *Biometrika* **39** 346–362.

Skellam, J. G. (1958a) On the derivation and applicability of Neyman's type A distribution. *Biometrika* **45** 32–36.

Skellam, J. G. (1958b) The mathematical foundations underlying the use of line transects in animal ecology. *Biometrics* **14** 385–400.

Smalley, I. J. (1966) Contraction crack networks in basalt flows. *Geological Magazine* **103(2)** 110–114.

Smith, G. E. J. (1979) Some aspects of the line transect method when the target population moves. *Biometrics* **35** 323–329.

Smythe, R. T. and Wierman, J. C. (1978) *First Passage Percolation on a Square Lattice*. Lecture Notes in Mathematics. **671** 1–196.

Snyder, W. V. (1978) Algorithm 531. Contour plotting. *ACM Transactions on Mathematical Software* **4** 291–294.

Sokal, R. R. and Oden, N. L. (1978) Spatial autocorrelation in biology. I. Methodology II. Some biological applications of evolutionary and ecological interest. *Biological Journal of the Linnean Society* **10** 199–228; 229–249.

Solomon, H. (1978) *Geometric Probability*. Society for Industrial and Applied Mathematics, Philadelphia.

Soneira, R. M. and Peebles, P. J. E. (1977) Is there evidence for a spatially homogeneous population of field galaxies? *Astrophysical Journal* **211** 1–15.

Southwood, T. R. E. (1978) *Ecological Methods*. Chapman and Hall, London.

Speed, T. P. (1978) Relations between models for spatial data, contingency tables and Markov fields on graphs. *Supplement Advances in Applied Probability* **10** 111–122.

Stimson, J. (1974) An analysis of the pattern of dispersion of the hermatypic coral *Pocillopora meandrina* var. *Nobilis* Verill. *Ecology* **55** 445–449.

Stiteler, W. M. and Patil, G. P. (1971) Variance-to-mean ratio and Morisita's index as measures of spatial pattern in ecological populations. *Statistical Ecology*. G. P. Patil, E. C. Pielou, and W. E. Waters, Eds. Pensylvania State University Press, College Town, Pa. I 423–459.

Strand, L. (1972) A model for stand growth. *IUFRO Third Conference Advisory Group of Forest Statisticians*. INRA, Paris. 207–216.

Strauss, D. J. (1975a) A model for clustering. *Biometrika* **63** 467–475.

Strauss, D. J. (1975b) Analyzing binary lattice data with the nearest-neighbor property. *Journal of Applied Probability* **12** 702–712.

Strauss, D. J. (1977) Clustering on coloured lattices. *Journal of Applied Probability* **14** 135–143.

Sukwong, S., Frayer, W. E., and Mogren, E. W. (1971) Generalized comparisons of the precision of fixed-radius and variable-radius plots for basal-area estimates. *Forestry Science* **17** 263–271.

Switzer, P. (1965) A random set process in the plane with a Markovian property. *Annals of Mathematical Statistics* **36** 1859–1863.

Switzer, P. (1967) Reconstructing patterns from sample data. *Annals of Mathematical Statistics* **38** 138–154.

Tallis, G. M. (1970) Estimating the distribution of spherical and elliptical bodies in conglomerates from plane sections. *Biometrics* **26** 87–103.

Tarrant, J. R. (1970) Comments on the use of trend surface analysis in the study of erosion surfaces. *Institute of British Geographers Transactions* **51** 221–222.

Taylor, L. R., Woiwod, I. P., and Perry, J. N. (1978) The density-dependence of spatial behaviour and the rarity of randomness. *Journal of Animal Ecology* **47** 383–406.

Thomas, M. (1949) A generalization of Poisson's binomial limit for use in ecology. *Biometrika* **36** 18–25.

Thompson, H. R. (1955) Spatial point processes, with applications to ecology. *Biometrika* **42** 102–115.

Thompson, H. R. (1956) Distribution of distance to nth neighbour in a population of randomly distributed individuals. *Ecology* **37** 391–394.

Thompson, H. R. (1958) The statistical study of plant distribution patterns using a grid of quadrats. *Australian Journal of Botany* **6** 321–342.

Tjøstheim, D. (1978) Statistical spatial series modelling. *Advances in Applied Probability* **10** 130–154.

Tobler, W. R. (1964) A polynomial representation of the Michigan population. *Papers of Michigan Academy of Science, Arts and Letters* **49** 445–452.

Tobler, W. R. (1969) The spectrum of US 40. *Papers of the Regional Science Association* **23** 45–52.

Tobler, W. R. (1973) Regional analysis: Time series extended to two dimensions. *Geographia Polonica* **25** 103–106.

Tobler, W. R. (1979) Smooth pyenophylactic interpolation for geographical regions. *Journal of the American Statistical Association* **74** 519–536.

Todd, C. D. (1978) Changes in the spatial pattern of an intertidal population of the nudibranch mollusc *Onchidoris muricata* in relation to life-cycle, mortality and environmental heterogeneity. *Journal of Animal Ecology* **47** 189–203.

Torelli, L. and Tomasi, P. (1977) Interpolation and trend analysis: Two geohydrological applications. *Mathematical Geology* **9** 529–542.

Torgerson, W. S. (1958) *Theory and Methods of Scaling*. Wiley, New York.

Trapp, J. S. and Rockaway, J. D. (1977) Trend-surface analysis as an aid to explanation for Mississippi valley-type ore deposits. *Mathematical Geology* **9** 393–408.

Tubbs, C. R. *The Buzzard*. David and Charles, Newton Abbott, Devon, England.

Tukey, J. W. (1970) Some further inputs. *Geostatistics*. D. F. Merriam, Ed. Plenum, New York. 163–174.

Underwood, E. E., de Wit, R., and Moore, G. A. (Eds.) (1976) *Proceedings of the Fourth Congress for Stereology, Gaithersburg, 1975*. National Bureau of Standards Special Publication **431**.

Unwin, D. J. (1970) Percentage RSS in trend surface analysis. *Area* **2** 25–28.

Unwin, D. J. (1973) Trials on trends. *Area* **5** 31–33.

Unwin, D. J. and Hepple, L. W. (1974) The statistical analysis of spatial series. *The Statistician* **23** 211–227.

Unwin, D. J. and Lewin, J. (1971) Some problems in the trend analysis of erosion surfaces. *Area* **3** 13–14.

Usher, M. B. (1969) The relation between mean square and block size in the analysis of similar patterns. *Journal of Ecology* **57** 505–514.

Usher, M. B. (1975) Analysis of pattern in real and artificial plant populations. *Journal of Ecology* **63** 569–586.

Veevers, A. and Boffey, T. B. (1975) On the existence of levelled beehive designs. *Biometrics* **31** 963–967.

Venter, R. H. (1976) A statistical approach to the calculation of coal reserves for the plains region of Alberta. *Canadian Mining and Metallurgical Bulletin* **69(771)** 49–52.

Vere-Jones, D. (1978) Space time correlations for microearthquakes—A pilot study. *Supplement Advances in Applied Probability* **10** 73–87.

Vere-Jones, D. (1980) Correction factors and stability of conditional intensity estimates. *Journal of Applied Probability*

Vincent, P. J., Haworth, J. M., Griffiths, J. G., and Collins, R. (1976a) The detection of randomness in plant patterns. *Journal of Biogeography* **3** 373–380.

Vincent, P., Sibley, D., Ebdon, D., and Charlton, B. (1976b) Methodology by example: Caution towards nearest neighbours. *Area* **8** 161–171.

Walden, A. R. (1972) Quantitative comparison of automatic contouring algorithms. *Kansas Oil Exploration Decision System technical report* 1–115, Kansas Geological Survey, University of Kansas, Lawrence, Kans.

Walker, J., Noy-Meir, I., Anderson, D. J., and Moore, R. M. (1972) Multiple pattern analysis of a woodland in south central Queensland. *Australian Journal of Botany* **20** 105–118.

Waloff, N. and Blackith, R. E. (1962) The growth and distribution of the mounds of *Lasius flavus* (Fabricus) (Hym: Foirmicidae) in Silwood Park, Berkshire. *Journal of Animal Ecology* **31** 421–437.

Walters, R. F. (1969) Contouring by machine: A users' guide. *Bulletin of the American Association of Petroleum Geologists* **53(11)** 2324–2340.

Ward, R. M. and Sprontz, W. (1976) Geographical analysis of drunken drivers. *Journal of Studies on Alcohol* **37** 997–1002.

Warren, W. G. (1972) Point processes in forestry. *Stochastic Point Processes*. P. A. W. Lewis, Ed. Wiley, New York. 801–816.

Watson, G. S. (1971a) Trend-surface analysis. *Mathematical Geology* **3** 215–226.

Watson, G. S. (1971b) Estimating functionals of particle size distribution. *Biometrika* **58** 483–490.

Watson, G. S. (1972) Trend surface analysis and spatial correlation. *Geological Society of America Special Paper* **146** 39–46.

Watson, G. S. (1975) Texture analysis. *Geological Society of America Memoirs* **142** 367–391.

Weibel, E. R. (1973) Stereological techniques for electron microscopy morphometry. *Principles and Techniques of Electron Microscopy*. M. A. Heyat, Ed. Van Nostrand–Reinhold, New York. **3** 237–296.

Weibel, E. R., Fisher, C., Gahm, J., and Schaefer, A. (1972) Current capabilities and limitations of available stereological techniques. *Journal of Microscopy* **95** 367–392.

Welsh, D. J. A. (1977) Percolation and related topics. *Science Progress* **64** 65–83.

Westman, W. E. (1975) Pattern and diversity in swamp and dune vegetation, North Stradbroke Island. *Australian Journal of Botany* **23** 339–354.

Westman, W. E. and Anderson, D. J. (1970) Pattern analysis of sclerophyll trees aggregated to different degrees. *Australian Journal of Botany* **18** 237–249.

Whitten, E. H. T. (1970) Orthogonal polynomial trend surfaces for irregularly spaced data. *Mathematical Geology* **2** 141–152; **3** 329–330.

Whitten, E. H. T. (1972) More on "irregularly spaced data and orthogonal polynomial trend surfaces." *Mathematical Geology* **4** 83.

Whitten, E. H. T. (1974a) Orthogonal-polynomial contoured trend-surface maps for irregularly spaced data. *Computer Applications* **1** 171–192.

Whitten, E. H. T. (1974b) Scale and directional field and analytical data for spatial variability studies. *Mathematical Geology* **6** 183–198.

Whitten, E. H. T.and Koellering, M. E. V. (1973) Spline-surface interpolation, spatial filtering and trend surfaces for geologically mapped variables. *Mathematical Geology* **5** 111–126.

Whitten, E. H. T. and Koellering, M. E. V. (1975) Computation of bicubic spline surfaces for irregularly spaced data. *Technical report* **3**, *Department of Geological Science, Northwestern University*.

Whittle, P. (1954) On stationary processes in the plane. *Biometrika* **41** 434–449.

Whittle, P. (1956) On the variation of yield variance with plot size. *Biometrika* **43** 337–343.

Whittle. P. (1962) Topographic correlation, power-law covariance functions and diffusion. *Biometrika* **49** 305–314.

Whittle, P. (1963a) Stochastic processes in several dimensions. *Bulletin of the International Statistical Institute* **40(1)** 974–994.

Whittle, P. (1963b) *Prediction and Regulation*. English Universities Press, London.

Wicksell, S. D. (1925) The corpuscle problem. A mathematical study of a biometric problem. *Biometrika* **17** 84–99.

Wicksell, S. D. (1926) The corpuscle problem. Second memoir. Case of ellipsoidal corpuscles. *Biometrika* **18** 151–172.

Williamson, G. B. (1975) Pattern and seral composition in an old-growth beech–maple forest. *Ecology* **56** 727–731.

Wilson, M. D. (1975) Comparison of fan-pass spatial filtering and polynomial surface-fitting models for numerical map analysis. *Geological Society of America Memoirs* **142** 351–366.

Yarranton, G. A. (1969) Pattern analysis by regression. *Ecology* **50** 390–395.

Zahl, S. (1974) Application of the S-method to the analysis of spatial pattern. *Biometrics* **30** 513–524.

Zahl, S. (1977) A comparison of three methods for the analysis of spatial pattern. *Biometrics* **33** 681–692.

Zubrzycki, S. (1957) On estimating gangue parameters. (In Polish.) *Zastosowania Matematyki* **3** 105–153.

Author Index

Page numbers in *italics* refer to lists

Subject Index